Optimizing Construction Management

Jozef Švajlenka • Terézia Pošiváková

Optimizing Construction Management

Sustainable Technologies, Waste Reduction, and Solutions for Wood-based Agricultural Structures

Springer

Jozef Švajlenka
Department of Construction Technology, Economy and Management, Faculty of Civil Engineering
Technical University of Košice
Košice, Slovakia

Terézia Pošiváková
Department of Breeding and Diseases of Game, Fish and Bees, Ecology and Cynology
University of Veterinary Medicine and Pharmacy of Košice
Košice, Slovakia

ISBN 978-3-031-84329-7 ISBN 978-3-031-84327-3 (eBook)
https://doi.org/10.1007/978-3-031-84327-3

This Springer imprint is published by the registered company Springer Nature Switzerland AG
The registered company address is: Gewerbestrasse 11, 6330 Cham, Switzerland

Acknowledgements

VEGA project 1/0228/24 “Research and innovation activities focused on efficient and sustainable wood-based construction technologies for agricultural construction”.

KEGA project 017TUKE-4/2024 “Implementation of an innovative interdisciplinary approach to knowledge sharing on agricultural structures”.

Introduction

Wood is one of nature's basic renewable materials. Knowledge of its basic properties is the first prerequisite for its proper use in various industries and in everyday life. Wood is the most versatile and widely used material (industry, construction, agriculture, everyday life). Due to its natural character, natural grain, and favourable physical properties, it is an increasingly sought-after environmental element. As wood is easy to work using cutting tools, it has become a means of expressing artistic ideas and human aesthetic sense (historical monuments made of wood, statues, altars, wooden buildings, churches, parts of castles).

Wood as the main load-bearing material for the construction of buildings based on wood is also becoming increasingly popular, thanks to the properties mentioned above. The benefits of wood mentioned above also produce the numerous advantages of modern constructions based on wood, including the fact that this is the so-called dry construction process, so the construction of prefabricated buildings is possible even in the winter months. The exclusion of wet processes also excludes failures and defects caused by the effects of technological moisture. When construction is complete, the construction based on wood can be used immediately, which means the investment starts to appreciate immediately. High-quality constructions based on wood have the ability to regulate interior humidity and can naturally maintain the optimum parameters of the indoor environment. Due to their sandwich construction design, they also have good acoustic and thermal-technical properties. They provide the necessary fire protection, thanks to their sandwich construction design, which is composed of layers of materials of different densities. The advantages of constructions based on wood with exposed wood structures include their aesthetic qualities. The natural texture, colour, and smell have a positive effect on the human senses. In addition to the fact that wood is the so-called renewable raw material, modern constructions based on wood offer significant environmental benefits compared to heavy ceramic and silicate constructions. Despite the undeniable advantages associated with the use of construction systems based on wood, their greater popularity is hampered by a low level of awareness among potential customers and users. Countries where constructions based on wood have a strong tradition and which are known for their active attitude towards innovation (Scandinavian countries) started to address this issue decades ago. The result was the creation of modern design solutions that have significantly contributed to increased construction efficiency.

The aim of this publication is to introduce the professional and scientific community to the field of wood construction and the use of structural systems based on wood for various types of construction solutions. The publication covers selected parts of the topic of wood used as a building material, wood constructions in the context of sustainability, construction systems of wood constructions and their application in various segments of building construction.

Thanks to the reviewers for the expert assessment: doc. Ing. Naďa Antošová, Ph.D. and Ing. Eva Wernerová, Ph.D.

Contents

List of Figures

List of Tables

Chapter 1
Construction Industry in Relation to Sustainability

In practically all areas of social and economic life, activities towards the concept of sustainable development can be observed today. According to Tsai and Chang (2012), sustainable development or sustainability as a concept and cultural definition has become synonymous with reducing environmental encroachment and degradation. However, the US Environmental Protection Agency uses the explanation provided by the Brundtland Report of 1987, which defines sustainability as "meeting the needs of the present without compromising the ability of future generations to meet their own needs" (Brundtland Report, 1987). This considerably broadens the definition of sustainability. The importance of sustainability is confirmed by the fact that the concept has also permeated regulations and legislation in the developed world. Sustainable development is legally defined, it is "development that preserves the possibility for present and future generations to satisfy their basic needs in life, while not reducing the diversity of nature and preserving the natural functions of ecosystems". This definition is based on a report published in 1987 by the United Nations' World Commission on Environment and Development (WCED). Sustainable development is understood as a purposeful, long-term (continuous), comprehensive and synergistic process, affecting all aspects of life (cultural, social, economic, environmental, and institutional) at all levels (local, regional, global), and leading towards a functional model of a particular community (local and regional community, country, international community) that satisfies the biological, material, spiritual, and social needs and interests of people, while eliminating or significantly limiting interventions that threaten, damage, or destroy the natural environment and life forms, does not burden land beyond a tolerable level, makes wise use of its resources and protects cultural and natural heritage (NSSD, 2001).

J. Švajlenka, T. Pošiváková, *Optimizing Construction Management*,
https://doi.org/10.1007/978-3-031-84327-3_1

1.1 Ecology as a Basic Pillar of Sustainability

Ecology is a fundamental pillar of sustainability (Hansmann et al., 2012), emphasizing the importance of our planet and the impact humans have on it. It involves recognizing our dependence on Earth's diverse ecosystems and taking tangible steps to minimize our impact (Isbell et al., 2017). This pillar prioritizes the protection of biodiversity, the balance and health of ecosystems, and ultimately, our own survival. Effective biodiversity management is crucial, as it highlights our interconnectedness with all life forms and the necessity to protect them (Xu et al., 2021).

In the context of sustainability, ecology plays a vital role in various aspects such as energy use, waste management, biodiversity conservation, and addressing emissions. Businesses are key players in protecting the environment by integrating environmentally friendly practices into their operations, contributing to climate change mitigation and promoting a circular economy to minimize waste and harmful discharge into ecosystems (Dangelico & Devashish, 2010).

Overall, ecology as a basic pillar of sustainability underscores the need to respect and preserve the natural world, recognizing that our well-being and survival are intricately linked to the health of our planet's ecosystems.

1.1.1 Ecological Criteria of Sustainability

The following section presents the basic ecological criteria of sustainability. These criteria are essential for assessing and ensuring environmental sustainability. Here are some key points extracted from the sources:

Ecological System Maintenance To ensure environmental sustainability, it is crucial to develop and maintain every element of the ecological system equitably. Neglecting environmental problems on a global scale can burden future generations with severe challenges, emphasizing the need for proactive measures to preserve nature and create a sustainable environment (Schlattmann et al., 2022).

The most frequently mentioned environmental problems on a global scale (Singh & Pradeep, 2017; Hardoy et al., 2024; Wernerová & Teichmann, 2017) are problems such as atmospheric pollution with chemical and solid substances, thermal pollution affecting water cycles in nature and changes in the structure of organisms, which leads to reduced productivity and stability of the landscape. Emissions of pollutants have a significant impact on morbidity, mortality, wear and tear of basic resources and agricultural production. Another important problem is the pollution of water by waste water from industry, human settlements, and agriculture, which has a negative impact on the hydrosphere (Salgot & Montserrat, 2018). These problems have global consequences and require complex solutions and international cooperation to mitigate them and protect the environment for future generations.

Environmental Monitoring A fundamental aspect of ensuring environmental sustainability is the competent assessment of the environment's state. This involves systematic observations at global, national, and local levels to evaluate the health of living and non-living natural objects, identify sources of impact, and predict changes to prevent negative environmental consequences (Bodini, 2012; Staudt et al., 2021).

Environmental monitoring and assessment of the state of the environment are fundamental aspects of ensuring environmental sustainability. These processes include the monitoring and evaluation of various components of the environment, such as air quality, water, soil, and biodiversity. Key competencies (Welford, 2016; Singh et al., 2019a) required to effectively manage an environmental management system (EMS) include:

- Knowledge of environmental regulations and laws
- Knowledge of EMS principles and standards
- Ability to identify, assess, and manage environmental risks
- Skill in environmental monitoring and measurement
- Ability to analyse environmental data and performance indicators
- Knowledge of emergency procedures and the ability to coordinate responses to emergency situations
- The ability to perform internal EMS audits and assessments

Resource Use and Waste Management Achieving environmental stability requires improving resource use quality, adopting safe technologies, implementing structural transformations in the economy, and managing industrial waste effectively. Transitioning to waste-free production technologies and minimizing harmful substances' impact on land, water, and air are critical for environmental sustainability (Nižetić et al., 2019).

Preservation of Natural Environment Countries are urged to preserve natural environments, landscapes, and biological diversity as part of the transition to environmental sustainability. This involves cleaning polluted territories, managing waste responsibly, preventing ecological emergencies, and implementing sustainable development programmes to maintain the natural balance and protect habitats (Grădinaru et al., 2020; Plati, 2019) (Fig. 1.1).

The ecological criteria for sustainability encompass maintaining ecological systems, monitoring environmental health, improving resource use, managing waste, and preserving natural environments. These criteria are essential for promoting a sustainable environment and ensuring the well-being of current and future generations.

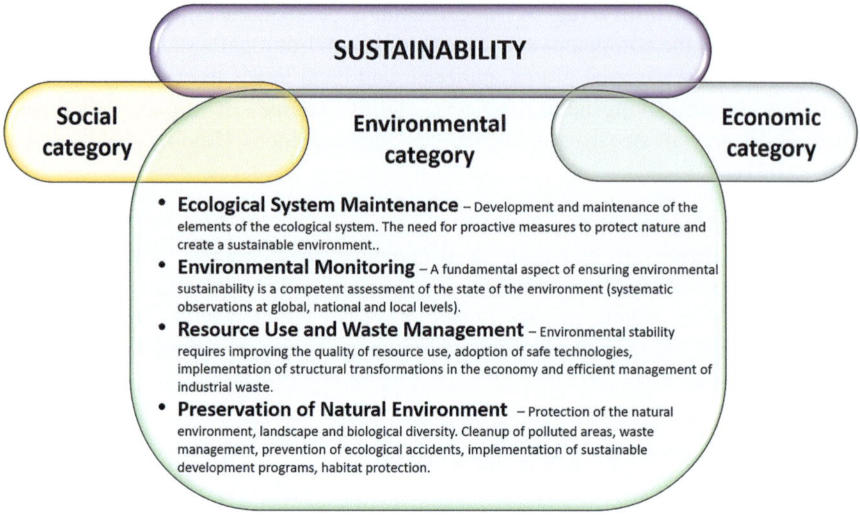

Fig. 1.1 Sustainability in connection with the criteria of the environmental field

1.1.2 Global Trends and Challenges Related to Sustainability

In connection with sustainability by the principles of ecology, the following global trends and challenges according to the EEA (European Environment Agency, 2015) are presented in five clusters. The first cluster, called social, includes: different global population trends, the increasing rate of urbanization in the world, the changing burden of diseases, and the risks of pandemics. The second technological cluster contains: accelerating technological progress. The third economic cluster includes: continued economic growth, multipolar world, more intense global competition for resources. In this context, the fourth and key cluster is environmental, containing: increasing pressure on ecosystems, increasing severity of problems and consequences of climate change, increasing environmental pollution. The fifth cluster contains: management and diversifying approaches to management.

The growing pressure on ecosystems is a serious problem not only in connection with climate change, but also with increasing environmental pollution, which threatens the stability and functioning of natural systems. The main factors (Malhi et al., 2020; Lu et al., 2018; Demková et al., 2023; Baranová et al., 2021; Smith et al., 2016; Cramer et al., 2018) exerting pressure on ecosystems and key variables include:

- Changes in land use and changes in land types, such as the conversion of arable land to grassland, which significantly affects the potential of regulatory ecosystem services.
- Increased emissions of pollutants into the air, which damage plants, slow down their growth and threaten entire ecosystems. An excessive amount of nitrogen leads to eutrophication of soil and water ecosystems.

- Population growth, increasing pressure on natural resources, rising living standards by increasing consumption and carbon footprint, waste, and invasive species.
- Increasing the average temperature, while individual seasons are warmer and extreme weather fluctuations are more frequent and intense.
- Changes in the water regime, such as prolonged periods of drought in the summer and autumn months associated with a lack of water, and at the same time an increased risk of floods.
- Negative impacts on ecosystems, biodiversity, and agriculture.
- Economic consequences, such as disruption of business operations, property damage, disruption of supply chains and infrastructure, which may ultimately lead to increased costs and higher prices of essential commodities.

In this context, the factories mentioned above and their consequences affecting ecosystems, human health, and biodiversity are often irreversible. Therefore, it is imperative that measures are taken to protect the environment and prevent further pollution in order to maintain a healthy and sustainable environment for current and future generations in the context of sustainability.

1.2 Sustainability in Relation to the Construction Industry

In line with technical as well as social development, according to Coale and Hoover (2015), Fang et al. (2016) and Wernerová et al. (2020), there has been a steady increase in population and uncontrolled extraction of raw materials for materials and energy production, along with the urbanization of the environment. The amount of harmful emissions and waste is also increasing, while raw material resources are limited. According to Azari and Abbasabadi (2018), just as the resources for the sustenance of a growing population (the problem of drinking water and food) are limited, so are the resources of raw materials needed for the construction of buildings and for the development of the infrastructure and industry needed to further secure people's living conditions. This also applies to common raw materials for the production of construction materials and products. For these reasons, the construction industry, according to Alwan et al. (2017), is generally considered to be one of the major contributors to environmental pollution. This gives rise to the need for changes in the approach to the design of building structures (Ďubek et al., 2021; Antošová et al., 2021; Ďubek et al., 2018). The result is, according to Nemet et al. (2016) and Mahmoud et al. (2019), that the concepts of green and sustainable building construction become one of the main aspects that can influence the solution to the global climate protection problem and contribute to the sustainable development of the planet. In terms of future prospects, it seems that the key factor of sustainable construction processes, according to Zokaei et al. (2013) and Chen et al. (2010), is not only the environmental impact of buildings during their life cycle, but also economic, social, and cultural aspects. Therefore, according to Hawken et al. (2010), the building industry must also evaluate sustainability from the perspective of both natural and human capital.

The evaluation of individual strategies, concepts, programmes, and activities in relation to sustainability can be carried out on the basis of principles (for managing people's activities) and criteria (for assessing the application of the principles). From the numerous principles and criteria set in the National Strategy for Sustainable Development for the SR (2002), the following can be applied to the construction industry:

- **Principle of promoting human resources development—criteria:**
  - ensuring the protection of human health
  - ensuring optimal development of human resources (in all life-enhancing areas)
- **Ecological principle—criteria:**
  - preservation and promotion of biodiversity and the vitality and resilience of ecosystems
  - optimizing the spatial arrangement and functional use of land and implementing its territorial system of ecological stability
  - preserving the high quality of environmental components—minimizing negative impacts on the environment
  - minimizing the use of non-renewable resources and favouring the use of renewable resources, but within the limits of their reproductive capacity
- **The principle of self-regulatory and self-sustaining development—criteria:**
  - identifying and exploiting natural simulated self-regulatory and self-sustaining natural mechanisms
  - promotion of closed cycles of production and consumption
- **Efficiency principle—criteria:**
  - preservation of optimal metabolic-energy cycles
  - minimization of raw material and energy inputs
  - reduction in the amount of outputs and minimization of losses
  - introduction and promotion of environmental economy tools
- **Reasonable sufficiency principle—criteria:**
  - reasonable and economical use of resources and their protection
  - promotion of appropriate forms of subsistence farming
- **Principle of precaution and foresight—criteria:**
  - prioritizing preventive measures over the elimination of undesirable consequences of activities
  - recognition of potential risks (including unverified risks)
- **Principle of respecting the needs and rights of future generations—criteria:**
  - preserving opportunities for future generations to use existing resources
  - preserving equal rights for future generations

These principles and criteria must be taken into account in line with the legislative requirements for constructions. Within the legislation of the Slovak Republic, the basic requirements for constructions according to Decree No. 532/2002 Coll. (NSSD SR, 2002) are characterized in the following sections: (I) Mechanical strength and stability, (II) Fire safety, (III) Hygiene, health, and environmental protection, (IV) Construction safety during use, (V) Protection against noise and vibrations, and (VI) Energy efficiency and thermal protection. Regulation (EU) No. 305/2011 of the European Parliament and of the Council (2011) incorporates a seventh requirement for constructions, i.e. (VII) Sustainable use of natural resources.

The philosophy of sustainability is fundamentally based on the three pillars, namely environmental, economic, and social. According to Hájek (2007), these areas represent the three basic areas of sustainable development, from which the basic criteria for sustainable construction are derived for the construction industry (Fig. 1.2).

According to Hájek et al. (2011a, 2011b, 2011c), the development of the construction industry in terms of sustainability will be determined by factors arising from the following general requirements of sustainable development:

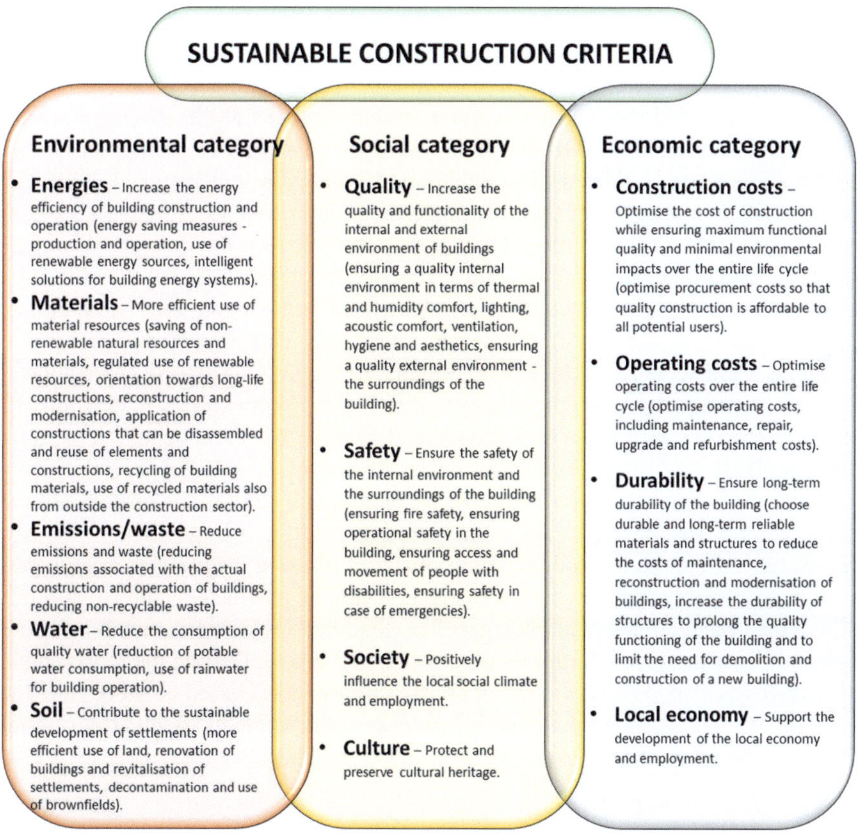

Fig. 1.2 Criteria of sustainable construction

- Reducing the overall intensity of buildings, especially in terms of energy consumption for construction, maintenance, and demolition of buildings, and materials recycling.
- Reducing the material intensity of buildings by using less energy-intensive materials in terms of sourcing and production, using renewable material bases, and using materials based on recycled raw materials.
- Assessing the impact of buildings on the surrounding environment in terms of environmental protection.
- Higher degree of industrialization of the construction industry due to more efficient and qualitatively better technological solutions.

In terms of sustainability, the construction industry has enormous potential to reduce the negative impact of its activities on the surrounding environment through the implementation of innovative and efficient solutions, which need to be assessed and quantified. Consequently, quantifying efficiency allows us to identify opportunities to implement more efficient, cleaner, and more sustainable technological solutions. Efficiency means achieving maximum outputs for a given level of resources used to complete a given process. According to Alchian and Demsetz (1972), measuring the efficiency of an organization means evaluating the relationship between the outputs it produces and the inputs it consumes. An efficient organization is one that produces maximum outputs for a given level of inputs, or that produces a certain level of outputs for a minimum level of inputs. Thus, according to Ruuska and Häkkinen (2016) and Werner and Richter (2007), efficiency means the maximum possible use of available resources regardless of their quantity and quality. Efficiency is related to productivity, but they are not identical. Although productivity is a measure of the final outcome, it describes, according to Giannetti et al. (2018) and Costanza et al. (2012), the ratio between the amount of outputs produced and one or more inputs used to produce them, regardless of the efficiency of their use, with efficiency adding an element of comparison with some known potential: an organization will thus be considered inefficient if, with a given level of inputs, it could actually produce more outputs. Efficiency is a diverse and synthetic concept, so measuring efficiency requires different ways of expressing it, using different criteria and indicators on the one hand, and the so-called benefit derived on the other hand.

Construction designs based on wood are undoubtedly part of the response to the trend towards sustainability. As a load-bearing construction material, wood is a renewable and environmentally friendly material offering many other advantages, such as dry construction, favourable thermal-technical parameters compared to traditional materials (brick, concrete), excellent load-bearing capacity to volume ratio, low emissions produced during its processing, and others. Despite the indisputable advantages associated with the use of construction systems based on wood, the low level of awareness among potential investors or users limits the popularity of constructions based on wood.

Chapter 2
Potential of Wood as a Construction Material

Wood is an organic material composed of cells. It is a composite made up of a chemical complex of cellulose, hemicellulose, lignin, and accompanying substances. According to Koželouh (2004a, 2004b), wood is anisotropic mainly due to the longitudinal shape of the wood cell and the orientation of the cell wall structures. In addition, anisotropy results from differences in cell size during the growth period and, in part, from the predominant direction of certain cell types (e.g. wood rays). The fine structure of the cell walls, the arrangement of cells in defect-free wood, and the growth irregularities of construction wood represent three structural stages, all of which have a considerable influence on the properties of wood as a construction material. For example, the submicroscopic structure of the cell walls explains, among other things, why drying and swelling perpendicular to the grain is generally 10–20 times greater than in the direction of the grain. The microscopic structure of defect-free wood means that wood is 20–40 times more rigid in the longitudinal direction than in the transverse direction. The macroscopic structure (e.g. bulges, fibre deflection) explains the tensile strength parallel to the grain, which can decrease from a value of more than 100 N/mm^2 for defect-free wood to <10 N/mm^2 for low-quality construction wood.

Wood of different species is made up of a set of plant tissues that differ in shape and function. The arrangement of the tissues is governed by certain laws that are related to the life (physiological) processes occurring in the growing tree. The morphology of the tissues and their arrangement into annual rings create macroscopic features that can be observed on a cross-section with the naked eye. The macroscopic features are very important for the practical differentiation of different wood species. These features provide essential information for the wood processor to identify the species of wood plant he is processing and to assess the physical, mechanical and technological properties of the wood, which determine its utility value.

J. Švajlenka, T. Pošiváková, *Optimizing Construction Management*,
https://doi.org/10.1007/978-3-031-84327-3_2

2.1 Basic Wood Directions and Sections

According to Čunderlík (2009), observing the physical and mechanical properties of wood, we find that the value of a measured property varies depending on the direction in which the property is observed. Defining wood directions is an essential condition for the correct use of wood. The property that indicates that wood has a different value of the property being observed in each anatomical direction (expressed in absolute terms in mm of swelling or in relative terms in %) is called *wood anisotropy*. Materials which do not have this property, i.e. the observed properties are independent of the direction of measurement (they are the same in each direction), are called *isotropic*. Three basic anatomical directions are defined for wood understood as an anisotropic material: longitudinal direction (l), transverse radial direction (r), transverse tangential direction (t):

- The *longitudinal direction* (axial) runs along the axis of the trunk. For small pieces of wood, it is determined by the longitudinal orientation of its fibrous structure, assuming that wood structure is made up of 70–90% fibrous strands that are oriented in the longitudinal direction. The annual rings, rays, and the perpendicular orientation of the longitudinal direction in relation to the tangential section help determine the longitudinal direction, especially of irregular pieces of wood.
- The *transverse direction* can be perpendicular to an annual ring or tangent to an annual ring. If the transverse direction is perpendicular to an annual ring, or is identical to a ray, the result is a radial direction. A precisely drawn radial direction intersects the annual rings at an angle of 90° and is perpendicular between two parallel tangential sections on a rectangular piece.
- The *tangential direction* (tangent) is drawn as a tangent to an annual ring. It intersects the rays at an angle of 90°.

According to Čunderlík (2009), by sawing a trunk at different angles in relation to its longitudinal axis (pith), or by sawing a trunk along its length in different positions in relation to the pith, we create section surfaces that differ in the macroscopic features shown. The set of these features on the section plane (surface) is called wood grain. There are three basic section planes, which, due to the longitudinal orientation of most of the cellular elements and the conically, even cylindrically layered growth of trunk wood, create distinctly different textures. The basic wood sections are: transverse section (T), longitudinal radial section (R), and longitudinal tangential section (T):

- The *transverse section* (frontal) is the first cut on the trunk that we make when sawing a tree. It is made perpendicular to the longitudinal axis of the trunk. The annual rings form a set of concentrically arranged rings with their centre in the biological centre of the trunk (pith). In the case of coniferous wood, the transverse section is made up of relatively well-defined lighter and darker circles. The springwood in circular-porous wood is made up of a visible, distinctly porous zone. Summerwood has less porosity, which is virtually invisible to the naked eye. It appears as a darker area. In the case of scattered-porous wood, the springwood and summerwood zones are virtually indistinguishable in terms of porosity and colour. It is just that the narrow layer of summerwood at the end of the annual growth is a darker colour.

- The *longitudinal section* runs along the axis of the trunk, in the direction of the fibrous structure of the wood. The longitudinal section may run through the *centre of the trunk*, through the pith, or at some distance from the pith, *as a tangent* to an annual ring. A *longitudinal section* running through the pith intersects annual rings at an angle of 90°. It runs in the direction of the trunk radius (radius), hence the name *radial section* or central section (R). The annual rings on a radial section form a set of parallel longitudinal lines and stripes. In coniferous wood and some deciduous wood, these longitudinal stripes vary more or less in colour, and in circular-porous deciduous wood, the longitudinal stripes have different grooving. A longitudinal section running tangent to the annual rings is called *tangential section* (T). Because of the conical structure of the trunk, the longitudinal section plane cuts through the annual growth at a slight angle in relation to the conical layers of the annual growth, which is reflected by the parabolic (fladrical) texture of the annual rings on this type of section.

2.2 Wood and Moisture Content

According to Koželouh (2004a, 2004b), the moisture content of wood is defined as the ratio of the mass of water removed (m_w) to the mass of dry wood (m_0). Dry mass is determined by drying in a kiln at 103 ± 2 °C. The moisture content of wood may be expressed as a fraction or as a percentage. Electric moisture meters are most commonly used to measure wood moisture content between 6% and 28%, as they are simple and quick to use. The accuracy of wood moisture meters is ±2%, which is sufficient for practical use in engineering. Currently, two principles are used: one principle is based on the measurement of moisture content using a dependent electrical DC resistance between two electrodes embedded in the wood; the other principle works with alternating current and determines the moisture-dependent dielectric properties of the wood in an electric field induced by two electrodes attached to the wood surface. Because wood is hygroscopic, it constantly exchanges moisture with the environment. For each combination of temperature and humidity of the ambient air, there is a corresponding moisture content level of the wood at which the diffusion of moisture entering the wood is in equilibrium with the diffusion of moisture leaving the wood. This moisture content is referred to as equilibrium moisture content. However, wood is rarely in a state of equilibrium moisture content because the climatic conditions of the environment constantly change. Moisture content and even the magnitude and rate of moisture movement have a major influence on almost all the properties of wood that are important for use in construction.

According to Kuklík (2005), due to the anisotropic structure of wood, there are significant differences in the drying and swelling of wood in the tangential and radial direction, which are the cause of the shape changes of wooden elements. According to Koželouh (2004a, 2004b), the internal stresses generated by anisotropic drying are primarily relieved by the formation of radial cracks. The larger the section and the faster the drying, the sooner the so-called desiccation cracks occur.

According to Koželouh (2004a, 2004b), the mechanical properties of wood also depend on moisture content. Moisture gain leads to a decrease in strength and rigidity values. This effect is partly explained by the swelling of the cell wall, which results in less cell wall material being available per area unit. More importantly, however, water enters the cell wall and weakens the hydrogen bonds that hold the cell wall together. Changes in moisture content above the saturation point of the fibres do not affect the mechanical properties since in this case only free water is stored in the cell cavities. The effect of moisture content changes on the different mechanical properties varies. For example, when compression stresses are applied parallel to the fibres, breakage is caused by fibre bending, with the moisture-sensitive hydrogen bonds playing an important role. On the other hand, tensile breakage parallel to the fibres means breakage of the covalent bonds as the microfibrils of the cell wall detach. Compressive strength is therefore more sensitive to moisture than tensile strength. For practical use, a linear dependence between moisture content of $8\% < \omega < 20\%$ and wood properties can be assumed.

2.3 Physical Properties of Wood

Determine the type of wood, we often need to use auxiliary properties of wood classified as physical properties. These include colour, gloss, density, and hardness.

2.3.1 Wood Colour

One of the first features that help to identify wood is its colour. This is not a matter of providing a precise physical definition of the colour of the wood. In assessing wood in terms of colour, it is necessary to consider the moisture content of the wood and the freshness of the cut. The moisture content of the wood can significantly alter its colour. In general, dry wood is always lighter than wood with a high water content. Similarly, exposure of the wood section being examined to atmospheric conditions can also alter its colour to varying degrees. An example of this would be a freshly sawn alder *(Alnus)* (light pink colour), which after a few days changes colour to bright orange as a result of oxidation of the accompanying substances by atmospheric oxygen. After alder wood is processed and dried, the colour changes back to pinkish red. Similar colour changes are observed in spruce *(Picea abies)*, pine

(Pinus sylvestris), and other species exposed to ultraviolet radiation (browning of wood). According to colour, wood is divided into these types:

- Pale wood (spruce, fir, hornbeam, and others)
- Brown wood (oak, ash, elm, walnut, and others)
- Red wood (pine, larch, holm oak, cherry, alder, and others)
- Green wood (acacia and others) Čunderlík (2009)

2.3.2 Wood Gloss

It is determined by the structure of the wood grain, and also by the section surface and the quality of wood processing. High-density wood has more gloss than porous, low-density wood. This rule is contradicted by hornbeam, which has a high density but is considered to be glossless. If we exclude the influence of the cutting tool, the most gloss is observed on longitudinal sections. In particular, the *rays* on the radial section *can significantly affect wood gloss (Acer, Quercus, Robinia, Ulmus, Cerasus,* and others). The highest porosity and thus the least wood gloss is observed on the transverse section. The possibility to assess wood gloss on products is significantly limited, as most wood products are treated with various types of coating Čunderlík (2009).

2.3.3 Smell of Wood

Smell is used in a *sensory assessment of wood*, especially when it is fresh or when it is processed (sawing, milling, sanding, etc.). The smell of wood depends on its chemical composition, especially on the content and quality of the accompanying substances. Terpene resins, tanning agents, oily compounds, and many other chemically complex substances that can be extracted from wood with water and polar and non-polar organic solvents have a characteristic smell. The content of these extractive substances tends to be higher in heartwood and lower in sapwood. Coniferous wood that has resin canals in its structure has a pleasant smell of resin. Fir *(Abies alba)* does not contain resin in its wood and has an unpleasant sour smell when fresh. *Quercus robur, Q. petraea, Robinia pseudoacacia, and Juglans regia* are examples of deciduous trees that have a characteristic terpene (bitter) smell. *Quercus cerris* (sour smell) can be distinguished from *Quercus robur* and *Q. petraea* by its smell when fresh. Fresh poplar wood (*Populus*) has an unpleasant smell. Birch wood (*Betula*) smells of mould. Linden wood (*Tilia*) retains the musty smell of old, unventilated rooms and buildings for quite a long time. The smell of wood is important when it comes to products that come into contact with food, as the food or foodstuff may be positively or negatively affected by the chemical composition of the wood. The smell of wood also has an effect on the users of wood constructions where wood is used to a certain extent Čunderlík (2009).

2.3.4 *Wood Density*

According to Koželouh 0, wood density is the most important physical characteristic of wood. Most of the mechanical properties of wood as well as the load-bearing capacity of joints positively correlate with density. According to Čunderlík (2009), it is according to wood density, i.e. mass and volume, that we judge the wood samples that we can hold in our hands. By rough estimation, we determine whether the wood is light, medium, or heavy. When relying on this physical property of wood, we must always consider the moisture content of the wood. According to density, wood is divided into these types:

- Very light (up to 400 kg/m^3), which includes fast-growing poplars
- Light wood (400–500 kg/m^3)—spruce, fir, pine, linden
- Moderately heavy (500–600 kg/m^3)—larch, willow, mountain maple
- Medium (600–700 kg/m^3)—birch, walnut, elm, ash, oak, beech
- Heavy (700 kg/m^3 and more)—hornbeam, acacia

The lightest wood, balsa (*Ochroma lagopus*), grows in the tropical forests of South America. Its density is about three times lower (120–150 kg/m^3) than that of spruce. The wood with the highest density, guaiacum (*Guaiacum officinale*), grows in Central America and the northern part of South America. Its dry density is about 1230 kg/m^3 Čunderlík (2009).

2.3.5 *Wood Hardness*

In macroscopic identification, we use wood hardness as a property that is closely related to wood density. The average hardness of wood (softwood or hardwood) or any significant difference in hardness (high degree of heterogeneity) between springwood and summerwood (coniferous wood) can be estimated by pressing a fingernail into the wood. Deciduous wood is less heterogeneous (less difference in density and hardness between springwood and summerwood) Čunderlík (2009).

2.4 Wood Defects

According to Čunderlík (2009) defects are defined as irregularities or variations in the texture and structure of wood or in the shape of the trunk or a certain wood product range which hamper or restrict its use for a given purpose. Such irregularities and variations generally degrade the physical and mechanical properties of wood and thus restrict its use for its standard purposes. We may also encounter variations and irregularities in the structure of wood which enhance the use of a product range. Such abnormalities in the structure of wood change its texture in particular, being sought after mainly for the production of ornamental veneers and valued as a special wood grain. However, the same variation or irregularity in the structure of wood may be regarded as a wood defect for some types of use and as a special wood grain for other types of use. Two types of defects are distinguished according to when they occur:

- Growth defects—occur during the growth of the tree
- Defects occurring during harvesting, handling, and storage of wood

Growth defects can be inherited (crookedness, convergence and forkiness of the trunk, sharp branching, etc.) or can be a reaction to abiotic (wind, snow, frost, etc.) and biotic factors (wood-destroying fungi, bacteria, viruses, insects, etc.). For example, from the perspective of wood processors, the most important raw wood defects are those most commonly encountered when purchasing wood logs. Based on the quality, the processor then decides what specific semi-finished product or product the log can be processed into. Similarly, the processor of raw wood must be able to assess how a defect will show in solid wood products once the wood is processed or how it will affect their quality.

2.5 Significant Properties of Wood as a Construction Material

The wood is one of the oldest and most widely used construction materials. Even in the present, wood is reclaiming the status it once had; thanks to its easy availability, favourable mechanical, structural, and thermal insulation properties, low bulk density, easy workability, aesthetic properties, recyclability, and low environmental burden.

In recent years, the Central European but also in many countries of the world with construction industry has been gradually returning to a more extensive use of wood. It is used not only for roof structures and the infill structures of all categories of residential buildings, but often as a load-bearing structural material or to complete the exterior and interior of various types of buildings.

In general, however, wood, as a renewable, widely available, as well as cost- and energy-efficient construction material, is still not used in contemporary architecture as extensively as it should be. In our country, the use of this material has a considerable tradition, but the key factor for a return to more extensive use of wood in our construction industry will be an increase in popularity among potential investors, architects, and designers, as, for example, with the help of new technologies, it is possible to apply wood construction systems in line with contemporary architectural trends.

According to Palfiová (2008), wood is ideal for the construction of load-bearing structures as well as for filling or cladding walls, ceilings, and floors. It is wood's unique properties, such as strength, elasticity, low bulk density, and easy workability, that make it suitable for such a wide range of uses. In proportion to its mass, wood is stronger than any other building material. A wood building weighs about one-tenth the weight of a comparable building made of concrete or burnt brick. If treated and cared for properly, it can withstand weathering and biological pests for centuries. This is evidenced by the well-preserved wooden buildings in the Nordic countries, but also in our country, which date back 200–300 years, and even 800 years in the case of some Viking buildings.

According to Štefko (2012), wood is originally a natural and organic material. The structure of wood shows that chemically, anatomically, morphologically, and macroscopically, it is an inhomogeneous and anisotropic biopolymer. Its structure mirrors the growth of the tree in a coded system of photosynthetic and subsequent biochemical reactions. This gives it a number of specific characteristics.

Physical and Mechanical Properties of Wood According to Štefko and Balent (2012), wood in building structures has excellent mechanical properties and a very good ratio of the load-bearing capacity of structural elements to their mass. With a suitably chosen structure and technology, which eliminates some of its negative properties, it enables the construction of cost-efficient load-bearing wall and frame structures, as well as the roofing of large spans.

Rheological Properties of Wood When wood is under a load over extended periods of time, its deformations increase and its strength decreases. The most significant change in deformation and strength occurs when periods of exposure to a load are extreme, i.e. when they are either very short or very long, for example, during the lifetime of a structure that holds roofing, snow, etc. Wood that is under a sustained load can be characterized as a visco-elastic, linearly behaving material. The structure of wood determines its deformation properties under a permanent load.

Wood Flammability/Fire Resistance Flammability is often wrongly used as an argument against the use of wood. Although wood does burn, it has a great advantage compared to other materials in that combustion can be accurately calculated (it contains 8–15% of water). For a tonne of wood to burn completely, 80–150 kg of water must evaporate from it, which requires a high energy input.

Impact of Wood on Health Natural wood contributes to the creation of a healthy indoor environment by virtue of the following:

- Wood has the ability to "breathe" (hygroscopic capacity of wood).
- Air is filtered when it passes through wood (detoxification effect).
- Active oxygen ions are not filtered out in the filtration process.
- It has a natural colouring, mostly consisting of brownish-yellow tones, which induce a feeling of warmth.

- Wood is a porous material and has a low heat conductivity (wood value from 0.11 to 0.21 W/m.K)—cladding and floors remain warm to the touch.
- Wood is an elastic material (floor construction systems), etc.

Aesthetic Properties of Wood Users find the colour and texture of natural wood pleasant. The use of wood in an artificial material environment allows us to achieve the appearance of surfaces that resembles the pleasant feel of the natural outdoor environment. It has a proven positive psychological impact on the user. Its technological properties allow it to be made into interesting structural details (exposed carpentry joints, sculptures, reliefs, polychromy, contrasting combinations with other materials, etc.). It is still an important tool of constructivism in architecture. Many buildings and wood structures by prominent architects, such as Frank Lloyd Wright, Alvar Aalto, Dušan Samuel Jurkovič, or by contemporary architects Imré Makovecz and Lubor Trubka, present the exceptional tectonics of wood by combining its exceptional mechanical and aesthetic properties (Štefko et al., 2010).

Technological Properties of Wood Wood is easy to work and join, its structure allows further processing into semi-finished products, and it also serves as a basic raw material for the production of large-size materials. Its good technological properties, together with excellent mechanical properties, enable the construction of cost-efficient and modern prefabricated buildings.

Štefko (2012) summarized the arguments in favour of using wood in construction as follows:

- Wood is a historical building material in the Slovak context.
- It is a strategic domestic renewable raw material.
- It has no equivalent in terms of its combination of properties, namely:
  - thermo-technical
  - aesthetic
  - environmental impact
  - utility
  - technological
  - mechanical
- Increasing the use of wood has a positive impact on the environment:
  - reduction in energy consumption
  - reduction in CO_2 emissions
- Favourable economic impact in the Slovak context:
  - added value
  - labour market
  - solutions to the current crisis

2.6 Wood-Based Materials Used for Wood Constructions

According to the shape of the transverse section, according to Kuklík (2005), wood for building structures is divided into these types:

- board timber (boards, planks)
- square timber (balks, battens)
- semi-finished timber
- cuts for construction purposes (struts, poles, etc.)

According to Kuklík (2005), wood-based materials are produced using various methods of industrial wood processing. The processing mostly involves the pressing (typically hot) of specially prepared wooden components obtained by mechanical splitting of wood. In the course of tree growth, inhomogeneous parts are formed in the wood material, for example, the so-called knots and other natural characteristics affecting the properties of the wood. The latest technological processes can eliminate these defects by selecting and then joining suitable parts of wood using various techniques (e.g. finger-jointing, agglomeration, layering). The following wood-based materials are mainly used:

Plywood Board This is mainly plywood, lath, and sandwich boards. **Plywood** is glued together using an odd number (at least three) of layers. In the construction industry, the most common type of plywood board is plywood, composed of layers of peeled or sliced veneers, so that the veneers of adjacent layers are usually at 90° angles.

Fibreboard This is made from fibrous sawmill waste. Wood fibres are processed using a wet or dry manufacturing process, layered and shaped using heat and pressure into the final product. Depending on the pressing pressure, temperature, type and amount of added substances, boards of different properties are produced. Fibreboard is the common name for seven different types of boards with significantly different properties. These boards are divided according to their density and production method (Table 2.1 by Kuklík, 2005).

Table 2.1 Types of fibreboard

	Board density		
Production process	Low <400 kg/m^3	Medium ≥ 400 kg/m^3, <900 kg/m^3	High ≥ 900 kg/m^3
Wet process	Soft fibreboard (SB)	Semi-hard fibreboard of low density (MBL)	Hard fibreboard (HB)
	Impregnated soft fibreboard (SB.I)	Semi-hard fibreboard (MBH)	Very hard fibreboard (HB.I)
Dry process		Medium density fibreboard (MDF)	

Among the above types of boards, dry process fibreboards (MDF) are growing in importance because their production is more environmentally friendly compared to the wet process (conventional fibreboards). In this production process, fibrous material is transported by airflow to the place where it forms layers to make a raw mattress. In order to achieve good bonding, the fibres are coated with various powdered synthetic adhesives, which make up about 10% of the weight of the board. The raw mattress is first pre-pressed to remove air. It is then cut to the required length and hot-pressed to make individual boards with a smooth surface on both sides. MDF boards are produced with a thickness of 1.8 mm to over 45 mm. Some MDF boards have been developed as vapour permeable in order to be suitable for roof and cladding construction.

Particleboard This is made from wood chips that are hot-pressed after the addition of glue. According to the method of production, two basic types of chipboard are distinguished:

- flat-pressed boards
- extrusion-moulded boards

In flat pressing, chips are layered and pressed between the heated plates of the press. In the case of extrusion-moulded boards, the pressing mixture from chips is forced under high pressure through an extrusion slit and formed into an endless strip. Globally, particleboard is produced in larger volumes than any other wood-based material.

Oriented Strand Board (OSB) It is made from very high-quality wood such as Scots pine. The flat chips are 50–75 mm long and their width is less than half their length. The outermost layers of the board contain flat chips oriented parallel to the longitudinal direction of the board, while the chips of the middle layer, which typically makes up to 50% of the volume, are arranged perpendicular to the longitudinal direction or randomly. The boards are normally bonded with phenol-formaldehyde resin, which makes up approximately 2.5% of the weight when used in powder form. The thickness of the board ranges from 6 to 25 mm. Good surface quality is achieved by grinding. OSB is now widely used in the construction industry because of its properties and its low bulk density compared to plywood and particleboard and its good workability.

Glued Laminated Timber Most oftenproduce elements of standardized dimensions—width 0.24 m, height 2.0 m, and length 32.0 m. These elements have excellent fire resistance. The charring rate of glued laminated timber is 0.5–0.7 mm/min without loss of load-bearing capacity. Therefore, glued laminated timber is favoured for constructions where large crowds gather—sports halls, exhibition halls, stadiums, etc. Because of its strength with low bulk density, it is also used for constructions whose elements have to be transported over greater distances. In order to increase the load-bearing capacity of glued laminated timber elements, it is also possible to reinforce them with strips of high-strength fibres with a thickness of approx. 2 mm. This modification will significantly increase the load-bearing capacity and the flexural rigidity of a wood element. However, the element must be carefully assessed in terms of its shear strength.

Laminated Wood It is a material similar to plywood where most or all of the veneers are glued with their grain oriented in parallel. In Europe, laminated wood is produced in Finland under the name Kerto. It is produced using 3.2 mm thick spruce veneers (3.0 mm thick when pressed), which are stacked lengthways with short overlaps in several layers and then pressed together. This produces elements with standardized dimensions: 26.0 m long, 1.80 m wide, and 21–75 mm thick. These elements can then be cut to smaller dimensions as required. The strength and rigidity parameters of laminated wood are highly favourable. The characteristic in-plane flexural strength of laminated wood is 51 MPa and the average value of the modulus of elasticity is 14,000 MPa. It is therefore possible to use laminated wood for the most demanding structures, such as roofing for halls and the like.

Densified Wood It is clear that the strength of a material increases with increasing bulk density. The low bulk density of wood is due to its structure. Therefore, if we want to significantly increase the bulk density of wood, we have to change its structure. One possibility is to press it to approximately half its volume at a temperature of 140–160 °C and a pressure of 10–15 MPa. The pressing speed is about 1 mm/min. This produces a new material called densified wood. Densified wood is most widely used in the form of densified plywood, which has a bulk density of around 1400 kg/m^3. This plywood is mostly used as thin spacers and inserts in exposed joints of wood structures.

Modified Wood A major disadvantage of wood is that, due to its organic origin, it is vulnerable to wood-destroying fungi and insects. To avoid this danger, wood must be kept permanently dry or prevented from being decomposed by the enzymes of wood-destroying fungi. If the hydroxyl groups (OH) in the molecular structure of wood are replaced with larger methyl groups (CH_3), it becomes highly resistant to wood-destroying fungi and will not swell or dry out. For example, beech wood treated by acetylation has far better durability. Modified wood is suitable for the production of windows, shutters, glued elements for bridge structures, etc.

However, it is only those types of materials which can retain their integrity and strength in the relevant moisture class throughout the expected life of the structures they are used for that are suitable for wood structures. The strength and rigidity parameters of wood-based materials and their other properties are determined by manufacturers on the basis of tests carried out in accordance with the relevant European standards.

Chapter 3
Structural and Connecting Elements of Wooden Structures

3.1 Basic Load-Bearing Structural Elements of Wood Constructions

The following section specifies specific structural elements used in the construction of wood constructions with their basic characteristics and properties.

KVH Structural Timber It is solid spruce structural timber (Fig. 3.1). Thanks to technical drying, they achieve excellent structural properties and do not contain any harmful substances. Finger joints allow KVH to be produced in standard lengths of 13 m and contribute to the high strength of each element. They are used, for example, for the frame structures of prefabricated constructions, gazebos, pergolas, roof trusses, and flat roof structures.

DUO/TRIO Solid Structural Timber It is glued from two or three centrally cut and length-jointed slats (Fig. 3.2). It is produced in standard lengths of 13 m. The production method contributes to the high strength of each element and provides excellent dimensional stability. It is mainly used for the frame structures of prefabricated constructions, roof and ceiling structures, gazebos and pergolas.

BSH Glued Laminated Timber It is glued laminated timber produced from technically dried spruce slats (Fig. 3.3). Defects with reduced strength are cut out. By eliminating several natural defects in the wood and layering the slats, we have a product that can meet high structural and architectural requirements. The timber can be straight or bent. They are mainly used for the construction of walls and ceilings, the construction of gazebos and pergolas, and also for bridge structures and structures requiring long-span beams.

J. Švajlenka, T. Pošiváková, *Optimizing Construction Management*,
https://doi.org/10.1007/978-3-031-84327-3_3

Fig. 3.1 KVH structural timber

Fig. 3.2 DUO/TRIO solid structural timber

Fig. 3.3 BSH glued laminated timber

CLT Panel CLT (cross-laminated timber) panels are a construction material with the best properties for modern wood constructions, made from layers of solid wood oriented perpendicular to adjacent layers and glued together (Fig. 3.4). This reduces deformations in the wood to a negligible level. This material is easy to build with. CLT buildings are environmentally friendly, precise, dimensionally stable, airtight, and diffusion open. They have excellent structural rigidity properties. The construction system is very simple and easy to manufacture, with a minimal number of layers. Thanks to their properties, CLT panels offer virtually unlimited possibilities in construction and architecture. They can be combined with other construction materials without complications. They are also an ideal construction element for

Fig. 3.4 CLT panel

Fig. 3.5 LVL structural laminated timber

passive houses. They are primarily used as internal and external walls, ceilings, and roofs in new family houses and apartment buildings. They are suitable for reconstructions, extensions, and indoor constructions. CLT panels form the supporting structure of a construction, whose external side requires thermal insulation. All common insulation materials such as polystyrene, mineral wool, and fibreboard are suitable for insulating CLT panels.

LVL Structural Laminated Timber It consists of several layers of pine and spruce veneer glued together, with a thickness of 3 mm (Fig. 3.5). Any defects in the wood are eliminated, resulting in a material with a uniform cross-section. The LVL composition ensures high strength and dimensional stability. Deformations and cracks caused by drying do not appear. The latest technology makes it possible to produce this material with a variety of dimensions, as it is produced with board lengths of 18 m and widths of 2.5 m. LVL elements are mainly used as: beams, rafters, columns, tie-beams, straining beams, window and door lintels, main beams, stiffeners, ceiling reinforcements, and various other industrial applications. The main advantages of LVL elements are: high strength, excellent dimensional stability, thin cross-section, light weight, minimization of structural settlement, and low moisture content.

Fig. 3.6 I-joists

I-joists Ceiling and wall joists for wood frame or solid wood structures with optimized thermal bridges (Fig. 3.6). There are three standard types with different flange widths, which differ in the height of the web—SJ45, SJ60, and SJ90. Standard lengths are 7, 9, 10, 11, 12, 13 m. Custom lengths can be produced according to the supplied documentation. Insulation of the web can be supplied. They are used as rafters, ceiling beams, or wall stands. The main advantages of these elements include: elimination of thermal bridges, dimensional stability, minimum dead weight, high load capacity of the beams, easy installation of the building equipment, processing using common woodworking tools, and availability of fasteners.

3.2 Wood Structure Joints

The structural elements of wood structures are usually joined using various mechanical fasteners (nails, staples, bolts, pins, screws and screw anchors, etc.), carpentry joints, and gluing.

According to Kuklík (2005), wood structure joints are divided according to their arrangement as follows:

- Setting, i.e. joining in the longitudinal direction
- Grouping, i.e. joining in the transverse direction
- Joining into joints, i.e. joining at different angles in the plane and in space

According to the nature of the action and type of fasteners according to Kuklík (2005), we distinguish the following joints:

- Flexible: carpentry joints, joints with mechanical fasteners
- Inflexible: glued joints

Mechanical Fasteners

According to Kuklík (2005), mechanical fasteners can be classified according to the transmission of forces in the joint as follows:

- **Pin-type fasteners**—These (nails, staples, bolts, pins, screws, etc.) are usually bent and pushed into the wood during the transmission of forces.
- **Surface-type fasteners**—These (screw anchors and nail plates) are inserted or pressed into the wood structural elements, and the transmission of forces thus occurs on the surface of the structural element.

The most commonly used steel fasteners are shown in Figs. 3.7, 3.8, 3.9, 3.10, 3.11, 3.12, 3.13, 3.14, and 3.15.

Carpentry Joints

According to Kuklík (2005), carpentry joints are among the oldest wood structure joints. Carpentry joints require high quality workmanship and their disadvantage is that they weaken the structural elements considerably compared to other joints and are labour-intensive to make. The load-bearing capacity of carpentry joints must always be demonstrated by structural rigidity calculations, taking into account the standard requirements and rules.

Fig. 3.7 Angled braces

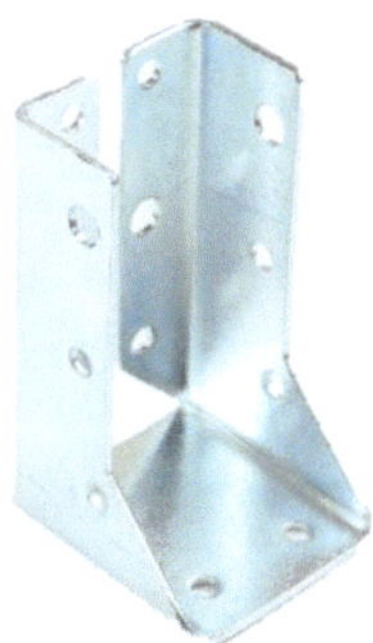

Fig. 3.8 Beam brackets

Fig. 3.9 Beam supports

Fig. 3.10 Beam braces and stops

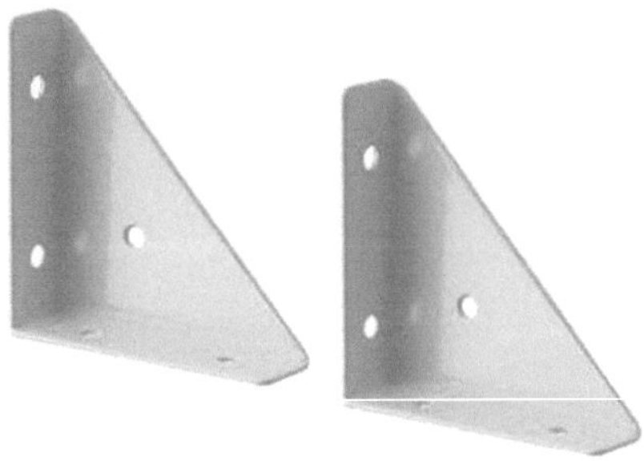

Fig. 3.11 Wall beam anchors

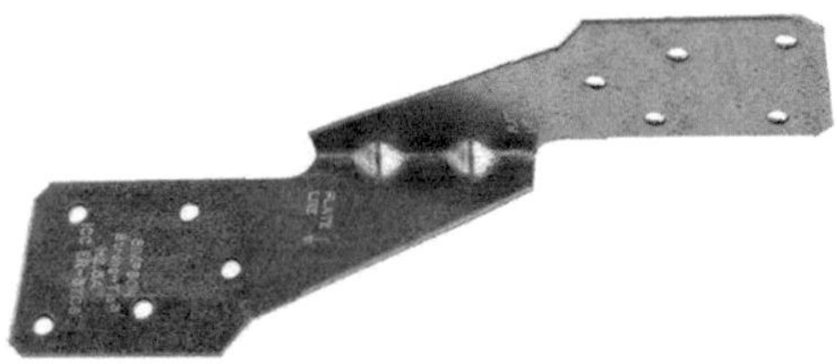

Fig. 3.12 Perforated plates and metal sheets

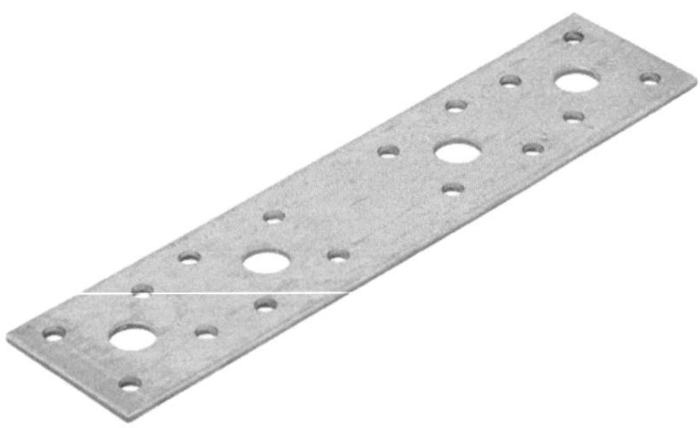

Fig. 3.13 Support brackets

Fig. 3.14 Universal braces and BULLDOG

Fig. 3.15 Bolts, nails, and rafters

Glued Joints

According to Kuklík (2005), glued joints are mainly seen in glued laminated timber. Glued laminated timber is glued using boards or planks with a maximum thickness of 45 mm, of various lengths, usually between 1.5 m and 5.0 m and 15 m, depending on the number of defects to be cut out of the timber, and thus on the amount of its shortening. Glued laminated timber typically features finger joints (Fig. 3.16), which join the individual pieces of timber, creating the so-called endless slat.

Fig. 3.16 Timber finger joint

Chapter 4
Construction Systems Based on Wood

The following types of construction systems for the construction of buildings based on wood are generally distinguished (Ramage et al., 2017; Allen & Iano, 2019):

- log construction system
- half-timbered construction system
- skeleton construction system
- column construction system
- panel construction system
- module construction system

The individual construction systems are specific in their design, especially of the load-bearing part, and specific demands are placed on them in terms of project preparation, production capacities, logistical support (sub-delivery of timber, warehouse management, transport, etc.) and assembly for on-site construction.

In terms of moisture (or water vapour) migration through the structure of a wood construction, there are two basic types of structures: diffusion-open and diffusion-closed structures. According to Lehocký (2018), a simplified description is that:

- **diffusion-open construction** = "breathing"—moisture passes through the structure, i.e. it migrates. Diffusion-open compositions typically do not contain a vapour barrier. However, it is still necessary maintain the order of the materials used so that those that are less vapour permeable start inside and those that are more vapour permeable should be closer to the exterior.
- **diffusion-closed composition** = "non-breathing"—prevents vapour from entering the structure. Again, moisture enters here, but in smaller quantities. This is most often the case in compositions where the final surface is a contact insulation with polystyrene and plaster. Inside the composition, there is wood and usually mineral wool. Steam passes easily through the mineral wool. It then reaches the polystyrene and passes through it with much more difficulty. It therefore stays longer in the wall structure and can condense into water under certain conditions.

J. Švajlenka, T. Pošiváková, *Optimizing Construction Management*,
https://doi.org/10.1007/978-3-031-84327-3_4

In order to prevent this phenomenon as much as possible, a vapour barrier is installed on the interior side. Steam passes through the vapour barrier in much smaller quantities and more slowly than through polystyrene. So the idea is that as little steam as possible gets into the structure and, if any does get in, that it manages to escape through the low-permeability polystyrene.

The degree to which the strengths of each construction system are understood and exploited is reflected in the price of the product and its delivery time, which is an important factor in most constructors' choice of a particular wood construction system.

4.1 Historical Context of Constructions Based on Wood

According to Štefko and Reinprecht (Štefko et al., 2009), the use of wood as a construction material in the distant past allowed people to produce valuable creations that not only served material needs, but also, thanks to the natural connection of wood with nature and its excellent properties, allowed them to complete their living environment and enrich it with elements that give people a sense of all-round well-being and aesthetic experience. The origins of wood constructions date back to prehistoric times, and over time, different construction systems have been developed and modified in different parts of the world. Research shows that the basic construction method and internal layout originates in the Early Stone Age: the emergence of a house with vertical walls and a structurally separated roof is thought to be a key milestone. Houses at this time were mostly made of wood and clay.

According to Kolb (2008) and Kuklík (2012), modern wood constructions are based on the original wooden buildings and structures, having undergone some development to meet the current requirements and needs. Until the first half of the nineteenth century, wood constructions were built using two basic methods, namely log and half-timbered construction, which was characterized by difficult carpentry joints. The origins of log construction are associated with Russia, from where the system spread to Sweden and Norway through Finland. The occurrence of log structures in Europe in the past was mainly linked to extensive forest wealth. Half-timbered buildings were common in areas where timber was not available in such large quantities. They gained favour in Eastern and Central Europe, England, Germany, Denmark, and the Netherlands. In different geographical areas, the buildings were characterized by different regional features. Houses of the late nineteenth and early twentieth centuries were characterized by balloon-frame construction, featuring lighter wood materials, mainly boards and planks. Another method, adopted around 1930, was platform-frame construction, using even shorter lengths of timber. This construction system is currently used in most American homes. By the mid-1970s, a change of perspective on an individual's position in nature and society led to a need to return to a more traditional and environmentally friendly way of life. Wood constructions have become an attractive housing alternative due to new innovative methods of construction and maintenance. The frame or column construction system, characterized by its simplicity of construction, economy, and architectural freedom has become a new trend. The bearing skeleton of frame structures consists of thin, standardized cross-sections of grown or glued timber, which are reinforced with sheathing of large-area material on both sides. Modern construction systems today make more optimal use of wood, i.e. specific construction systems based on column assemblies only use a fraction of the wood used by their predecessors. Wood is only used for the load-bearing part of a construction and the other elements combine different material bases for the best possible structural performance (Fig. 4.1 prepared by Kolb, 2008).

According to the Association of Wood Processors of the SR (2020), the share of construction of wood buildings has increased from about 2% to almost 10% in the last 10 years. Compared to other countries, however, this is the lowest share (Scandinavia 70%, Germany 50%, Austria 45%, Czech Republic 20%, the USA 65%, Canada 80%, England 40%).

4.2 Log Construction System

According to Štefko and Reinprecht (2004), Scandinavia can be seen as the birthplace of log constructions. The medieval experience of shipbuilding was translated into the construction of strong and tight log buildings. The Vikings traded with distant lands and brought back home not only the items they received in exchange, but also various techniques and knowledge that they sold to people in remote parts of Scandinavia. Norwegian craftsmen developed a construction style that interlocked horizontally stacked crossbars with interlocking joints. This construction system, called "laft", is one of the most sophisticated log cabin technologies. It was made using mostly firs, whose tips were cut off 2 years before felling to draw resin into the wood. The status of log cabin technology is still preserved in Norway today:

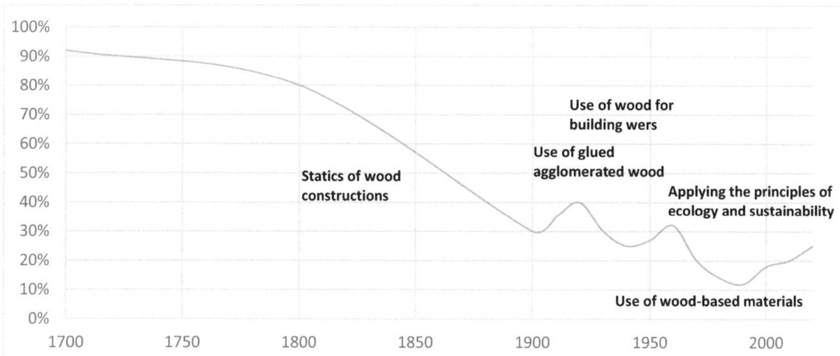

Fig. 4.1 Historical development of the share of wood in constructions based on wood

the rural landscape is dotted with old log constructions, many of them several years old. In North America, the first log buildings can be found from the first half of the seventeenth century. This log house construction craft was brought by newly arrived Scandinavians who came to settle on the east coast of what is now the United States. Later, these technologies began to blend with the influences brought to the "new world" by German immigrants. We now probably owe the expansion of log cabin construction on this continent to the Germans, along with new immigrants from Scotland and Ireland. Wherever timber was plentiful, the building of log dwellings was economically viable. The reason that few log buildings survive today is their original makeshift nature. The original log cabins were very primitive and only had earthen floors with their walls resting directly on the ground, which was responsible for their short lifetime. In the 1970s, there was a revival of this old craft in North America and since then the industry has experienced an incredible boom. With the emergence of new materials and designs, the technical parameters of log cabins and their utility value have also increased tremendously. Log cabins are certainly not among the cheapest structures available on the market today, and their complexity and sophistication is unbelievable in this day and age. Despite the progress of modern technology, the principle of building craft log buildings has remained virtually unchanged. It still requires extensive knowledge of the materials and structure, and a robust understanding of its strengths and weaknesses (Mojprojekt, 2018).

Log buildings have exterior walls made of worked tree trunks, i.e. logs of different profiles such as prisms or logs. The contact points of both exterior and interior walls are interconnected with carpentry joints to ensure the overall rigidity of the structure. In traditional log buildings, no additional thermal insulation is used in the exterior walls, which means the exterior solid logs also perform a thermal insulation function. The joints between the individual logs are traditionally filled with a material based on synthetic or natural thermal insulating fibres.

In terms of production technology, log construction systems can be classified as:

1. genuine log constructions:
   (a) traditional log constructions from solid wood
   (b) modern log structures from solid or glued wood
2. false log constructions

The characteristic features of log construction systems, according to Kolb (2008) and Čuláková (2013) are: high level of craftsmanship, special choice of wood, high wood consumption, fixed layout, volumetric and shape changes of the construction, and the settling of the construction. Log constructions react differently to variations in relative humidity in the interior. Wood has a strong ability to absorb and release moisture depending on the relative humidity of the air in which it is located, and therefore the interior climate in a log cabin constantly reacts naturally, without the need for mechanical treatment of the air. In terms of architectural and urban design rules, these types of constructions are limited to areas tolerant of these types of constructions.

Fig. 4.2 Example of a traditional log construction

4.2.1 Traditional Log Constructions from Solid Wood

Log constructions (Fig. 4.2) are the continuation of traditional folk architecture of the Central Europe, especially in the mountainous parts where wood was abundant. The development of log structures in the world is marked by migration (CEDER, 2018). Log constructions are built by joining exterior and cross beams with carpenter's joints, which ensures the rigidity of the resulting structure. The so-called traditional logs or partially squared logs are used for construction. The thickness of beams depends on the load on the wall, and on the climate in the case of residential buildings; the dimensions of beams are usually 18/20–20/22 cm. The joints in the loading joints are filled with thermal insulation material, which is either exposed or covered with a cladding batten. As standard, neither the interior nor the exterior side is sheathed or plastered, so the wood remains as exposed.

4.2.2 Modern Log Constructions from Solid or Glued Wood

Modern log constructions they are designed and constructed in line with the thermal-technical legislative requirements, i.e. higher standard criteria for the thermal-technical quality of constructions, making use of the broad technological possibilities in industrial wood production, applying construction research and development of building materials (thermal insulation, cladding), etc. A modern log cabin consists of structural elements from solid wood or glued profile wood.

Modern Log Constructions from Solid Wood

The basic structural element of modern solid wood log constructions (Fig. 4.3) is machine-profiled wood. A solid log element typically consists of one piece of timber or is joined lengthwise using different types of joints. Log elements may have

Fig. 4.3 Example of a modern log construction

Fig. 4.4 Example of a modern log construction from glued wood

the maximum moisture content of a freshly felled tree, or may be partially dried (naturally, artificially), or may be completely dried for the climatic region in which they will be used for construction. Exterior walls are typically built without additional insulation, but compositions containing insulation (external, internal, or inside the structure) are also built, depending on the required thermal-technical parameters of the resulting building (Drevodomgerlach, 2018).

Modern Log Constructions from Glued Wood

Log constructions from glued wood (Fig. 4.4) are also called Scandinavian-type log constructions. According to Hemström et al. (2011) and Wimmers (2017), Scandinavian-type wood constructions or modern log constructions originate in the Nordic countries, where they enjoy a centuries-long tradition. Thanks to their popularity, they are currently spreading to other European countries. According to Allen and Iano (2019) and Wacker (2010), sandwich Scandinavian-type wood constructions are almost exclusively saddle-roofed, with an arrangement that achieves a high housing standard on a relatively small built-up area and with interiors exhibiting high quality workmanship. According to Nilsen (2021), these houses can be characterized as functional and understated. According to Harte (2017), wooden vertical or horizontal boards with tongues and grooves, which are ventilated, are used for these houses almost universally. Glued log elements are made by gluing together multiple parts for the desired thickness, height, or length of the element. The resulting construction elements are tailored to the required dimensions and profiles for application in constructions. The profile can be square, rectangular, or round. Glued log elements are more difficult to manufacture and more expensive than solid log elements. However, they have their advantages, such as better quality (due to the possibility to select better quality parts of timber), dimensional stability of the element, aesthetics of the log cabin without cracks on the exterior or the interior side, improved the thermal-technical properties of the log wall (the entire thickness of the element without cracks is involved in thermal resistance), etc. Plasterboard elements are often fitted to the internal walls and ceilings, and the so-called fix-o-pan boards, similar to chipboard, with tongues and grooves, covered in paintable and washable foil, are also used. These boards allow quick and easy installation and their use also offers benefits for the user.

4.2.3 False Log Constructions

An analogue of the above-mentioned variants of log constructions, according to Varga (2024), is their imitation in the form of the so-called false log constructions. The main motivation to build this type of construction is the visual aspect of the building, i.e. the aesthetic aspect with different static and thermal properties of the construction. Such a construction may be clad on the exterior side with a thinner log structure (imitating a log cabin), while the load-bearing part of the construction may be built using a column, masonry, or reinforced concrete structure. According to the thermal-technical requirements of the resulting construction, the sandwich structure is fitted with thermal insulation of the required thickness.

4.3 Half-Timbered Construction System

The origin of the half-timbered construction system (Fig. 4.5) dates back to ancient Egypt, ancient Greece, the Roman Empire, and ancient China. As one of the earliest building processes, half-timbered houses were located in places where timber had to be saved. The basic component of the earliest buildings were beams, and the space between them was filled with various materials (most often clay). The floor was made of clay and the roof was a simple structure typically covered with reeds or boards. Such constructions were not durable as the columns were gradually saturated with moisture and gradually fell apart. Later, frame supports on a stone cushion sunk into the ground were used, and the walls were also reinforced with straw,

Fig. 4.5 Example of a half-timbered construction

branches, or reeds. Over time, the development of crafts (especially shipbuilding) made it possible to produce new structural parts, beams, and stands with progressive rectangular cross-sections. The experience of generations of builders showed that inclined frame sections (struts) significantly improved the stability of the construction. Half-timbered houses gradually spread throughout Europe, including the Scandinavian countries, the Netherlands, Switzerland, and France. In each country, the technology changed and adapted to local conditions (e.g. in England the walls were additionally insulated with sheep's wool). Stone foundations and logs did not appear until the fifteenth century. Half-timbered houses were a common feature of the urban landscape of the fifteenth and sixteenth century, especially in northern Europe—in Great Britain, Germany, and Poland (Drevoastavby, 2013; CMD, 2014).

According to Kohout et al. (1996), the essence of a half-timbered construction is a timber frame typically filled with masonry. Vertical load is transferred through columns and shear strength is ensured by the action of the frame and the masonry. The skeleton consists of groundsills, columns, struts, trusses, and lintels. The timber frame is exposed and is an architectural feature of the structure, which places increased demands on its workmanship and finish.

4.4 Skeleton Construction System

The skeleton system is a spatial load-bearing system and is made up of vertical (columns) and horizontal (spans) load-bearing elements (Fig. 4.6) (Estav 2023). The filling of the exterior wall and the internal partitions are non-load bearing. To ensure spatial stability, stabilizing reinforcing elements are placed in the structure to increase the strength of the walls and, if necessary, in the ceilings (if the strength of a ceiling structure based on large-area materials is not sufficient).

Fig. 4.6 Example of a skeleton construction

The skeletal structural system is most commonly used for the construction of public, school, and residential buildings. According to Kuzma (2017), the spatial load-bearing system allows building larger room partitions, improving the visual and aesthetic qualities of the construction. According to Čuláková (2013), the system allows considerable compositional freedom, high flexibility of the layout solution, while the load-bearing skeleton and the exterior walls enclosing the space remain mutually independent, and the wooden skeleton can be visible according to the requirements, but also covered on both sides with the chosen surface treatment. The use of heavy machinery is required for building skeleton constructions, and there is increased labour intensity in completing the details of the structural parts of the building. In terms of technology, it is also a disadvantage that most of the elements are made by gluing slats and structural units, which ultimately contributes to higher energy consumption of the final product. Closely related to this fact is the directly proportional increase in the amount of bound energy as well as the resulting higher emissions in the production of the final product.

4.5 Column Construction System

The column construction system (Fig. 4.7) is widespread, especially in the USA and Canada, and is also known as the "two by four" (2 × 4 inches). In the original American "two by four" system, according to Koželouh (2004a, 2004b), 50 × 200 mm boards are used for horizontal lintels, soffits and rafters, with an axial distance of the bearing elements of 400 or 600 mm. The column construction system is almost exclusively completed on the construction site, where it does not require much mechanization. The frame structure is built on a concrete base with foundations, or a foundation on earth screws is used. Spatial stability and rigidity is

Fig. 4.7 Example of a column construction

provided by walls sheathed with large-area materials or reinforced with diagonals embedded in the wall columns. The construction system is based on 50–60/100–140 mm profile columns. If the profile is not suitable from a structural point of view, the necessary profiles are created by grouping or by creating columns with a segmented or composite cross-section.

If column constructions need to meet stricter thermal insulation requirements applying to passive houses, columns made of grown timber must be replaced by thin-walled "ladder beams". Column constructions can be divided into two subgroups according to the construction system:

- The so-called balloon-frame, where columns run from the foundation sill (fixed into the foundation) to the eaves and the floor beams are fixed—attached to the columns from the sides.
- The so-called platform-frame, where columns are interrupted by the ceiling and the ceiling joists are laid on the top block of the frame.

The column construction system allows for certain spatial variability of the architectural design of the building, i.e. changes even during construction without a significant increase in the cost of construction. The disadvantages of the column construction system are directly associated with the technological characteristics since the construction of a column construction takes place directly on the construction site, which is especially a disadvantage of larger construction projects, as the entire construction process and the construction itself are exposed to adverse weather conditions, which can ultimately negatively affect the future properties of the resulting construction. This construction system requires higher efficiency and quality of work on the construction site. A significant disadvantage of the resulting construction is its low heat accumulation capacity; therefore, various accumulation elements are used in the construction to ensure better thermal comfort during use (Čuláková, 2013).

4.6 Panel Construction System

One of the most widely used construction systems for the construction of buildings based on wood is the panel construction system. This is in fact a modified column system, where panels are manufactured in a factory off-site and, once assembled, are transported to the construction site, where the final construction is assembled. According to Liška (2012), prefabricated wood constructions are based on three basic panel design principles. The most common panel design principles seen on the market are: wood-framed panels, structural insulated panels (SIPs), and layered all-wood panels. They are arranged according to the history of their design, the first being the wood-framed panel. The oldest design principle, the wood-framed panel, was patented in 1880. Wood-framed panel constructions became globally popular immediately after their launch. The popularity of panel wood constructions was also helped by the international armed conflicts in the nineteenth century, giving rise to the need to quickly build and dismantle field hospitals and accommodation facilities.

The panel construction system is based on the production of individual types panels (floor, exterior, partition, ceiling, gable, roof) in a production hall, with transport to the construction site and rapid assembly using machinery. Wood panels can be classified according to several criteria:

By size:

- small-size—small-format
- large-size—full-wall

According to the construction purpose:

- floor
- exterior
- partition
- ceiling
- gable
- roof

According to the stage of completion in production:

- thick panels—first stage of prefabrication (without hole fillers)
- final panels—second stage of prefabrication (with hole fillers)
- final panels—third stage of prefabrication (with final finishes)

4.6.1 Small-Format Prefabricated System: γ-ha Prefabricated Blocks

According to the company Andante (2023) and Drevostavby-car (2023), the construction system from prefabricated blocks (γ-ha) is suitable for all types of buildings where the intention is to achieve significant energy savings for heating the building. It is particularly advantageous in the construction of residential buildings, i.e. family houses, apartment buildings, and especially roof superstructures of apartment buildings, where the following are particularly appreciated: low load on the building, easy handling and vertical transport, dry assembly process, speed of assembly. The system allows easy disassembly without damaging the parts and thus their reuse for another building. As the parts are relatively light, their assembly and handling is also possible without the use of heavy lifting equipment. The system has significantly fewer thermal bridges across the surface of exterior walls. The construction system consists of the following construction elements: foundation sills, exterior wall and partition modules, lintels and wreaths. Foundation sills is the starting elements of the construction system. It is anchored directly into the foundation slab (Fig. 4.8). 150-mm module thickness is used for internal partitions or for sheathing for liapor concrete walls. 300-mm thickness is for the exterior walls of low-energy buildings, 450-mm thickness is for the exterior walls of passive buildings. Assembly is dry and the construction system can be disassembled. Lintels are used to bridge construction openings for windows in the exterior walls and for doors in partitions. Wreaths complete the wall or partition assembly. The precision of the modules' production and the ingenious arrangement of the individual elements allow the system to be structurally variable, rigid, and stable, while the entire assembly is dry and very simple. The internal space of the modules is filled with thermal insulation material, with two basic options: a) thermal insulation is inserted at the production plant, in which case it is recommended to use materials that are compressible with shape memory, such as mineral or glass wool; b) thermal insulation is inserted after the walls and partitions have been assembled on site, in which case blown cellulose, wood fibre, cork chippings, etc. are used.

Fig. 4.8 Example of the installation of individual small-format modules

The weight of the largest piece without thermal insulation is about 15 kg, which allows comfortable handling of modules.

4.6.2 *Small-Format Prefabricated System: STEKO Prefabricated Blocks*

This is a "Swiss" building system from dried solid spruce wood called the STEKO construction system (Fig. 4.9). According to Kleskeň (2019) and Bosák (2016), the basic element of this construction system is a wood module, manufactured in Switzerland. It is the so-called wood "brick" made in a factory with a building module—a multiple of 8 cm. The length dimensions are 16 cm, 32 cm, and 64 cm and the height is 24 or 32 cm. The width (building thickness) of the block is 12 and 16 cm. This guarantees the precision of the products and also the accuracy of on-site assembly. The modules are very easy to connect side by side and on top of each other (much like Lego pieces), forming the building structures that make up the walls and partitions of the house. The originality of the system lies in the fact that the modules are joined (interlocked like a brick building) without any gluing. The dry assembly allows very fast and precise construction according to the laying plan (processed in the factory). The plan precisely marks and numbers the blocks for a particular wall and places them onto pallets so that there can be no confusion and improvisation in the assembly of a particular wall. The individual wall modules are connected by means of the special shape of their top and bottom side (tongue and groove) and additional wooden "dowels" (made of poplar wood). The system components are delivered from the factory on pallets. After the guide sills are anchored to the foundation, the STEKO modules are stacked easily, quickly, piece by piece, without the need for expensive machinery and cranes. Without the use of mortar, cement or glue, a quality house is built in 2 or 3 days. The wall and the partitions anchored into it, built in the same way, form a structurally compact unit with a solid connection. The ceiling elements are placed on the STEKO wall closed with a "wooden" reinforcing wreath into which the truss is anchored. All types of truss systems can then be used for the truss. These "wooden brick" walls can be filled

Fig. 4.9 Installation of individual STEKO modules

with different types of thermal insulation with a thickness of 12 cm (partitions) and 14 cm—exterior walls + thickness of external insulation (depending on whether the house is insulated with a contact insulation system or other insulation system). This makes it possible to build to the "passive standard" with smaller construction thickness. Compared to the usual method of wood-based construction (panels), STEKO walls do not need a vapour barrier. This ensures natural temperature balancing in both summer and winter. The pores and cells of the solid wood regulate temperature and humidity—much like the human skin does. The solid STEKO wall thus serves as a heat reservoir in winter and keeps the interior pleasantly cool in summer.

4.6.3 Small-Format Prefabricated System: Brikawood Prefabricated Blocks

The creator of this wood-based system is the French startup Catharhome, which created the ecological and economically efficient Brikawood wood bricks (2020) (from English words brick and wood). The Catharhome system (Fig. 4.10) is based on the "lego" principle, i.e. the assembly of individual components without the use of glue, screws, and nails. Each wooden element, i.e. the brick, consists of two sides which form the exposed layer of both interior and exterior surfaces. The system uses two main edges with projections and grooves that slide into each other. The wooden bricks are divided into basic 50-cm long bricks, corner bricks, and half 25-cm long bricks. The Brikawood system withstands the weather well, thanks to the agglomerated wood elements. The wood-based construction elements are non-toxic and easily recyclable.

Fig. 4.10 Brikawood construction elements

4.6.4 Small-Format Panel Systems: SIP

One of the most commonly used small-format wood panel system technologies is the SIP (Structural Insulation Panel) technology (Fig. 4.11). According to Liška (2012), this construction system was developed in the USA, when it was first used by architect Alden P. Dow for residential construction in 1952 in the state of Michigan. From the 1950s to the 1970s, the production of SIP panels in the USA experienced mixed success and was slowly incorporated into building legislation. With the emergence of OSB boards in 1981, the modern structural insulated panel was created, and since 1995 it has been used on the Slovak and Czech market. The basic element of this construction system is an industrially manufactured panel, which is produced by gluing planks of large wood-based materials (OSB boards) to a core, which forms thermal insulation. The resulting panels have the required mechanical and physical properties at low weight and are very easy to process on site. The cores of the panels are made of EPS-F expanded flame retardant polystyrene, which is generally known to be good thermal insulation (Tejchman, 2014).

This is a construction system that, according to Stasiak-Betlejewska (2016), is used to produce low-energy and passive buildings. Currently, this system has been developed to the extent that it can compete with frame structures and prefabricated wood structures, as it exhibits good thermal insulation properties, high strength and minimal material costs compared to the rest of the market. Panels are used in the construction of external walls, partitions, ceilings, and roofs, where the thickness of the insulation determines the required thermal insulation properties. SIPS panels can be produced as large elements with ready-made openings for windows and doors. These large-size elements are transported to the construction site using special semi-trailers and are handled by a crane. The advantage of this system, according to Cox (2018) and Meng et al. (2018), is quick assembly on site. The disadvantage of large panels, according to Murillo et al. (2019), is the more complex manufacturing technology and more difficult storage, transportation, and handling of parts in the factory and on site. Another problem may occur if the

Fig. 4.11 Example—small-size panel systems

foundation plate does not meet the prescribed dimensions or if the investor requires additional changes to the project.

4.6.5 Large-Size Full-Wall Sandwich Panels

This system originated in the Scandinavian countries and is used extensively for wood-based construction in Germany and Austria. The basis of the construction of large-scale panels is a timber-frame structure forming a timber panel frame adapted to the function it will perform in the construction. The individual panels are adapted with the thickness of the load-bearing parts and the composition of the sandwich structure layers to the functional requirements (exterior, partition, floor, or roof panels) (Knut & NES BAU, 2020) (Figs. 4.12 and 4.13). The load-bearing wood-frame is most often sheathed with large-area materials (e.g. OSB). For the panel frame, spruce or fir wood with a cross-section of 60/140, 70/180 mm, dried with 15% residual moisture, or glued laminated wood (KVH beams) is usually used. The space inside the panel's bearing skeleton is filled with thermal insulation, which also serves as sound insulation. The panels are finished in different stages of completion, i.e. either just the frame sheathed on one side, the frame sheathed on

Fig. 4.12 Production hall for full-wall sandwich panels

Fig. 4.13 Example of a panel construction

both sides with thermal insulation or with hole fillers, or a panel with surface finishes.

The main advantage of the construction of the panel system according to Kolb (2008) and Čuláková (2013) is the possibility of maximum preparation of the construction parts (prefabrication) of the building in the factory without the influence of weather conditions on the production process, and quick assembly and completion of the building on the construction site, which takes about 3 months in standard conditions. Thus, the prefabrication and finishing rate of the panels is the key point of efficiency improvement. By using automated operations, the quality of production increases in direct proportion and, with repetitive execution, workers acquire craftsmanship skills, increasing efficiency. The panel construction system enables the use of modern production process lines with automation nodes and the efficient use of truck load capacity. The panel construction system utilizes, as an integral part of fast and efficient assembly, cranes capable of seamlessly handling the construction of the building even in more challenging terrain conditions, ultimately reducing construction time on site.

4.6.6 Construction System from Solid Wood Panels

Modern methods of construction based on wood are highly diverse and can be individually customized and combined. According to Kujanová (2010), current trends in wood-based construction increasingly use various innovative methods of industrial production of wood-based elements.

Among these constructions, according to Singh et al. (2019b), are laminated solid timber systems (e.g. cross-laminated timber (CLT)) and buildings made of assembled cross-sections. Solid wood structures usually have the form of large-size sheet elements. According to Kobl (2011), depending on the manufacturer or the product, solid wood structures can be single-layered or multi-layered—in the latter type, the multiple layers are either glued together or joined with pins or nails. Construction elements joined with pins do not require an adhesive and use the effect of a lower moisture content. Wooden pins with a moisture content of 6% are inserted into drilled holes with a moisture content of 12%. According to Veselý (2013), the moisture content in the pins and the moisture content in the structure will soon equalize, causing the inserted pins to expand, which ensures a firm joining. Other systems use pressure-inserted aluminium pins. The layers are typically glued together with PVAc (polyvinyl acetate) adhesives or PUR (polyurethane) adhesives.

The interior side of sheet elements from solid wood can be lined, left as a finished interior surface, or fitted with a slatted frame to house the building's utility connections. As is the case with any external wall composition, it is necessary, according to Berglund and Burgert (2018), to align the diffusion resistance of the interior layers with the diffusion resistance of the exterior layers. The advantage of leaving the interior solid wood surface as a finished interior surface is that it can bind excessive interior air moisture and release it again, helping to create the optimal humidity of the interior. External wall layers do not normally require plastic layers with diffusion resistance. Cross-laminated timber systems are, according to Quesada-Pineda et al. (2018), air-tight by their very nature. Other systems, such as skeletal structures or structures with layers joined with pins, usually require the installation of additional layers to prevent air ingress by lining the walls with boards based on wood (e.g. OSB) or other air-tight layers.

Coniferous wood is used as the material (spruce wood, fir wood). According to Woodard and Milner (2016), certain less common materials (particle boards, OSB boards) are also be used. According to Takabayashi et al. (2019), wall elements are manufactured with CNC-milled holes and grooves for utility connections. Ceilings and roof elements can either be manufactured in the same way as wall elements, or wall elements are combined with other structures. The degree of prefabrication or readiness for assembly can vary. Individual sheet elements are assembled in the superstructure at the construction site according to the assembly procedure of the specific manufacturer.

Cross-laminated timber (CLT) panels are a construction material with properties suitable for modern wood constructions (Fig. 4.14), made from layers of solid wood oriented perpendicular to adjacent layers and glued together. According to Brandner et al. (2016), this reduces deformations in the wood to a negligible level. The material makes it easy to build virtually all types of residential and non-residential buildings. According to Espinoza et al. (2016), CLT constructions are eco-friendly, precise, dimensionally stable, air-tight, and diffusely open, but can also be diffusely closed. They have excellent statics-related properties. The construction system is very simple and easy to manufacture, with a minimal number of layers. According to Lukacs et al. (2019), thanks to their properties, CLT panels offer virtually unlimited possibilities in construction and architecture. They can be combined with other construction materials without complications. They are also a suitable construction element for passive houses. They are primarily used as internal and external walls, ceilings, and roofs in new family houses and apartment buildings. According to Popovski and Gavric (2016), they are suitable for reconstructions, extensions, and indoor constructions. CLT panels form the supporting structure of a construction, whose external side requires thermal insulation. All common thermal insulation materials are suitable for insulating CLT panels, including polystyrene, mineral

Fig. 4.14 Example of a panel construction from solid wood panels

wool and other, environmentally more suitable alternatives, for example, those based on wood fibre or other natural materials.

It is mainly prefab products and products from solid wood, which include cross-laminated wood products, that make a modern wood construction, according to Brandner et al. (2016) and Flores et al. (2018), an interesting and sustainable construction technology.

4.7 Module Construction System

According to Juřenčák (2019), module buildings gained popularity at the beginning of the twentieth century. The post-war period brought an even greater boom in module construction, but the main material of these houses was still steel and concrete. Raising standards and updating building codes in the 1990s brought more environmentally friendly alternatives to module houses to the fore, attracting new potential customers to the market. The pursuit of sustainable living, new technological practices, innovative wood-based materials, and fast-paced lifestyles have all contributed to the increased interest of society in timber buildings made of large prefabricated modules.

Wood-based module constructions are characterized by a high degree of prefabrication. The whole building is fabricated in the production plant and transported to the destination. On site, the cells can be made into an object by both vertical and horizontal adjustment. A tractor-trailer is required to transport the modules. Due to their not insignificant weight (sometimes several tonnes), a relatively powerful crane is needed for unloading, which is used to unload them directly from the transport vehicles to the storage site. Both groups of machines must be on site throughout the assembly period, as the crane must hold each module for the entire time it is being fitted and stabilized. Depending on the size of the building, they are usually brought in individually on a transport vehicle. Cellular structures are mainly used for temporary buildings, but also for residential and civil construction. In the conditions of the Slovak Republic, the module construction system based on wood is not as widespread as the similar system of module buildings with steel skeleton construction.

Analysis of the current state of the art in wood-based construction has shown the potential for increased construction efficiency in terms of speed of construction, production, or operating costs and quality.

4.8 Special Types of Constructions Based on Wood: Multi-Storey Buildings and Halls

Load-bearing structures can be defined according to Hájek et al. (2011a, 2011b) as spatial parts of objects whose mass is shaped and arranged in space so as to achieve an effective interaction in the transmission of external forces acting on the building.

Different structural systems can then be distinguished according to the nature of the mass distribution of the load-bearing structure. The choice of a particular structural system for a particular building depends on the functional requirements, the production possibilities, and on the overall economic viability. There are two basic groups of civil engineering buildings, with associated groups of structural systems: multi-storey buildings and halls.

4.8.1 Multi-Storey Buildings

For multi-storey buildings (and also for many single-storey buildings), according to Hájek et al. (2011a, 2011b) the key elements of a structural system are vertical load-bearing structures (walls, columns, pillars), and the key elements for halls are horizontal load-bearing structures (roof structures). For the construction of high-rise wood constructions based on wood, various types of glued timber elements and large-area glued elements (CLT, X-LAM, etc.) are typically used.

Basic types of structural systems for single- and multi-storey buildings or wall-structural system are:

- Longitudinal
- Transverse

A special group of structural systems are prefabricated space unit (module) systems, which are essentially technological variants of wall or column systems which are divided as follows:

- Wall structural system
  - longitudinal
  - transverse
- Column structural system
  - with longitudinal frames
  - with transverse frames
  - with two-way frames
  - with board ceilings
  - with concealed bridging or capitals
- Combined structural system
  - with transverse walls and transverse frames
  - core system
- Prefabricated space unit (module) system of wood structure

Multi-Storey Buildings and Height Limits

According to Véghová (2018), wood constructions are increasingly popular in the world, and the trend is not limited to cottages, wood houses, and family houses using elements made of wood. It also includes office buildings, non-residential premises, and high-rise buildings.

According to Sandanus (2010), multi-storey wood constructions are a promising area for the use of wood, as they require much less energy to produce compared to "traditional" materials. The vast majority of low-energy or passive constructions have a wooden load-bearing structure. In Western European countries, wood is commonly used as a load-bearing material for multi-storey buildings. The height of multi-storey buildings is usually limited only by fire resistance requirements. Taking into account the mechanical properties of wood and ensuring the correct dimensioning of the elements, it is not difficult to set accurate height limits.

According to Véghová (2018), wood has been underused as a building material due to the restrictions imposed in laws on wood inspection, but also due to the fact that it was silicate and steel bases that were used extensively before 1990. The slow and ineffective development of fire safety regulations at that time was also an obstacle. It was therefore much easier to tighten the rules for wood constructions. The current state of technology shows that wood construction has progressed far beyond log houses and portable cabins. The approaches of EU countries to the fire safety of wood constructions are therefore under constant review.

Until recently, according to Véghová (2018), two storeys were accepted in different countries for residential and accommodation buildings, four storeys for non-production buildings and three storeys for production buildings. However, these limits also significantly limited the use of buildings due to the requirements for the highest permissible fire risk. The fire resistance requirements for this type of structural design were simply too strict. The most significant change was the possibility to build 5-storey timber buildings if certain conditions were met. The design and construction requirements for wood constructions are based on Eurocode 5. What also began to be taken into account was whether the structural elements of wood constructions were protected or not. The aim of the revision was not to mindlessly increase the previously permitted number of storeys or height of wood constructions, instead, it categorized it according to the degree of protection. This mainly means covering the wooden parts of structures with non-combustible protective cladding and, if present on the structures, filling the cavities with non-combustible materials. For practical purposes, a distinction must be made between *exposed* and *protected* wood.

- In the case of exposed wood, the required fire safety is achieved by applying strict height restrictions. In the case of *protected wood*, it is necessary that it does not ignite or smoulder inside the structure under the protective cladding during the required period of fire resistance. In addition, the problem of fire spreading through the interior of wood structures must also be addressed if structural cavities are unfilled or filled with combustible materials. In the case of multi-storey wood constructions, all cavities must now be filled with non-combustible materials that can withstand temperatures of 1000 °C and cannot drop or fall out of their cavities. If the cavities of wood constructions are not treated in this way, their maximum permitted number of storeys is reduced.

- In the case of *exposed wood*, where the filling of cavities does not meet the above requirements, the construction is a combustible structure and has a limited maximum permitted height. For non-production buildings, the limit is 9 m of the level of the last storey above ground level, for residential and accommodation building, the limit is three storeys, and for production buildings, it is four storeys for protected wood without the required cavity filling or three storeys for exposed wood. The fire resistance of load-bearing and fire-dividing structures in these cases shall be 60–90 min. In addition to fire-resistant structures, it is important to have suitable fire exit routes. Multi-storey buildings have longer fire exit routes. Protected and partially protected routes must be created to provide the required protection to escaping persons for a sufficiently long period of time. The specific type depends on the type of use of the construction, the number of people, and the distance to safety. The latter is determined by the number of storeys and their area.

Multi-storey wood constructions, which have a long tradition in Scandinavia and North America, still face prejudice in this country. According to ASB (2019), investors are mostly concerned about the fire protection of such buildings. However, wood construction structures may have a significantly high level fire resistance (up to REI 120). While in our country we achieved a height of five storeys a year ago, the first wooden high-rise buildings are already being erected abroad. Among the most famous are the projects for a 40-storey (133 m) building in Stockholm (C.F. Møller) and an 80-storey building in London with a height of 300 m (Oakwood Timber Tower, PLP Architects). According to experts, the fire safety of a well-designed and executed wood construction is no worse than that of a well-designed and executed masonry building.

Another example of multi-storey construction based on wood according to Sandanus (2010) is from the Italian city of L'Aquila, where a tender to build several dozen apartments after the 2009 earthquake was won by contractors building apartment buildings with cross-laminated wood load-bearing structures due to the speed of construction, good fire resistance, and good resistance to seismic loads.

4.8.2 Halls

Glued laminated timber-frame structures are, according to Hájek et al. (2011a, 2011b), due to their relatively low weight and high load-bearing capacity, mainly used for the construction of large-span hall structures for sports buildings, exhibition halls, shopping, and storage halls and halls for light industry. Glued frame structures can be designed for spans of 40, 60, and 100 m. The limitation is mainly due to the handling of parts produced in a factory and their transport and assembly on site. The advantage of glued laminated structures is their greater resistance to wood-destroying pests and moisture. At the same time, it is possible to achieve high precision which is maintained during the life of the building due to the greater dimensional stability of glued elements when exposed to moisture changes compared to conventional timber. The frames of glued laminated timber halls can be three-jointed with joints at the foot and top or two-jointed with joints only at the foot of the frame. The shape of the frame may be arched, semicircular, hockey stick type, or with a gable or hipped roof shape. The spacing of the frames is generally 4–6 m. The roofing is made of longitudinally oriented glued tie-beams fixed by means of contact sheets between tie-beams and rafters or complete wooden roof panels with internal thermal insulation. The production technology of glued laminated structures allows a custom design of structure shapes, with the entire production of structures taking place in the factory in an industrial way and the assembly to be completed on site.

Variants of glued laminated timber hall frames:

- Two-jointed frame
  - glued laminated timber element
  - joint in the support
- Three-jointed frame.
  - glued laminated timber element
  - joint in the support
  - joint in the top

Sports halls, wooden churches, or wooden parts of other buildings are landmarks of many cities. The use of glued laminated timber opens up new dimensions for its applications. It makes it possible to create buildings of larger spans without the use of supports, to shape the beams in various ways and thus to complete even more demanding constructions.

Chapter 5
Analysis of Alternative Material Solutions of Structural Parts of Construction Systems in the Sector of Agricultural Buildings from the Point of View of Sustainability

The interest in sustainable development in the last 40 years has shaped our current sustainability approaches, such as the definition of sustainable development, i.e. development that satisfies the needs of the present without affecting the ability of the generations to come to satisfy their own needs (Attia, 2018; Kishawy et al., 2018).

Sustainability consists of economic, environmental, and social factors and their relations (Kamali & Hewage, 2017). Social priorities involve a growing global consensus on which we should protect and improve animal welfare (Lin & Shih, 2018).

Sustainable agriculture is agriculture that satisfies the needs of the present without affecting the ability of the generations to come to satisfy their own needs. Sustainability has become a central concept partly due to the apparent need to produce sufficient food for the ever-growing world population and to simultaneously preserve the environment and contain climate change (Moustafa, 2016). Therefore, the focus has been on food production, environmental services and social goods although we also need to consider their consumption, along with other factors, for example, losses and waste generated during production, sale, and distribution (Arcari, 2017; Veissier et al., 2008; Langford & Stott, 2012).

Sustainable intensification is understood as increasing the volume of production, while concurrently improving efficiency in the use of inputs and reducing harmful impact of food production on the environment (Koc & Ceylan, 2012). Social and economic changes are needed to understand what outputs are required of those involved in food production, along with new research to deal with objectives which are more complex than simply achieving larger production volumes (Opp et al., 2018).

Agriculture and its various fields have undergone numerous transformations and developments, improving its quality. Rural areas and areas outside city centers (peripheries), which have been neglected up until recently, are now gaining an importance (Sanye-Mengual et al., 2018). It is mainly these areas' significance in terms of ecology and sustainability that is growing—they are significant as a place

J. Švajlenka, T. Pošiváková, *Optimizing Construction Management*,
https://doi.org/10.1007/978-3-031-84327-3_5

where the environment's natural elements exist and recover, acting as a basic precondition of preserving human health, including healthy and quality food and nourishment. Food production is an important social factor that is crucial for human survival, so its qualitative aspect is a key determinant of living standards and is also important in terms of sustainability (Stevens et al., 2018; Drozd & Leśniak, 2018).

At present, the emphasis is put on the development of environmentally friendly technologies. It is not only about energy-saving technologies, but also an overall approach that aims to minimize any undesirable environmental impacts of plant operation, or the production and use of products themselves. It is important to always choose a technology that is more environmentally friendly and to consider all its consequences (Hong et al., 2018). The incentives for experts to start dealing with eco-vacancy are mainly population growth, limited fossil stock, and rising environmental pollution (Kristjansdottir et al., 2018). The aim is to optimize their parameters to minimize the burden on the environment, in particular with regard to resource use; production, transport, and distribution; consumption; and end-of-life disposal (Filho et al., 2018; Do & Cetin, 2018).

The most important aspect in assessing environmental impacts is energy consumption (Ďubek et al., 2018; Invidiata et al., 2018), which is related to a limited amount of energy sources as well as accompanying negative phenomena such as atmospheric emissions contributing to acid rain, smog, and greenhouse effect (Embodied energy (EE), Global warming potential (GWP), and Acidification potential (AP)) (Kamali & Hewage, 2016; Tulevech et al., 2018). In particular, the greenhouse effect is most likely the most serious of the current environmental impact at all.

Currently, the Life Cycle Assessment (LCA) analysis method is applied worldwide. The LCA method has a fixed structure and is performed according to the international standards of the ISO 14040 series. Commercially available process databases and material and energy flows are used to efficiently process LCA studies (Østergaard et al., 2018; Lu et al., 2017). It is one of the most important information tools for environmentally oriented product policy. The LCA can be defined as collecting and evaluating inputs, outputs, and possible environmental impacts of the product system throughout the life cycle (Marique & Rossi, 2018). According to ISO 14040 and 14,044, life cycle assessment is performed in four different phases.

The subject of the analysis was the most frequently used construction systems of agricultural buildings considered in terms of selected aspects of sustainability. In the course of this analysis, we focused on the evaluation of the so-called cradle-to-gate evaluated variants. "From Cradle to Gate" is the assessment of the partial life cycle of a product from the extraction of raw materials ("cradle") to the factory (exit) gate (i.e. before it is transported to the consumer). The use and disposal phases of the product are omitted in this case. Cradle-to-gate evaluations are sometimes the basis for environmental product statement (s). In particular, the following aspects were considered in the evaluation: EE, GWP, and AP.

5.1 Subject of Investigation and Methodology of Analysis

The research object was real buildings designed for agricultural primary production (livestock farming). The present analysis of the environmental advantages of a modern prefabricated wooden panel construction with traditional building systems is based on prefabricated reinforced concrete components and a steel structural system through the above parameters. To evaluate the sustainability of the compared construction systems, a part of the LCA was implemented within the set boundaries of the assessment.

5.1.1 Research Subject

The subject of the analysis was selected structural systems of industrial buildings used mainly as agricultural objects in the V4 countries (Czech Republic, Slovak Republic, Poland, and Hungary—Central Europe) (Sýkora et al., 1992). Examined and compared construction systems were designed on the basis of different construction materials (concrete, steel, and wood). All of the construction variants being compared were implemented by the so-called fly-by-assembly (based on similar assembly principles with a different material basis). In addition, the compared objects had a comparable capacity of objects in terms of the built-up area and the obsolete space (Table 5.1), while maintaining the comparable thermal and technical properties of the structure.

5.1.2 Methodology of Environmental Parameters Assessment

The environmental evaluation of construction systems was made by LCA methodology in mode "Cradle-to-Gate". The materials were investigated in three constructions groups: foundation, superstructure, and roofing. Considered building materials and structures are shown and specified in the notes below (Figs. 5.1, 5.2, 5.3, 5.4, 5.5, and 5.6 prepared by Sýkora et al., 1992). The composition of floor structures

Table 5.1 Structural characteristics of construction systems

Type	1 Module			All construction		
	Size	Built-up area (m^2)	Built-in space (m^3)	Size	Built-up area (m^2)	*n*
C1a	6 × 12	72	324.54	12 × 72	864	3894.48
C1b	6 × 12	72	295.20	12 × 72	864	3542.40
S1a	6 × 12	72	303.36	12 × 72	864	3640.32
S1b	6 × 12	72	295.74	12 × 72	864	3548.88
W1a	6 × 12	72	293.88	12 × 72	864	3526.56
W1b	6 × 12	72	252.00	12 × 72	864	3024.00

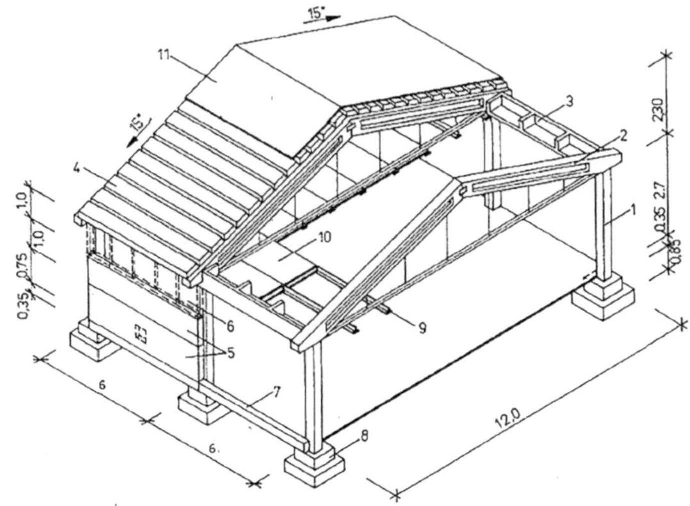

Fig. 5.1 Structural and materials characteristics of construction system C1a (Sýkora et al., 1992). Note: 1, column 0.3 m × 0.3 m; 2, three-armed beams with steel rod; 3, ribbing wreath; 4, roof rack plate SZD; 5, peripheral sandwich panel rough 0.2 m; 6, window belt; 7, base reinforcement 0.2 m × 0.35 m; 8, prefabricated foundation foot; 9, beam ceiling (wood T profile); 10, ceiling panel (thickness 0.1 m); 11, covering (fiber–cement boards)

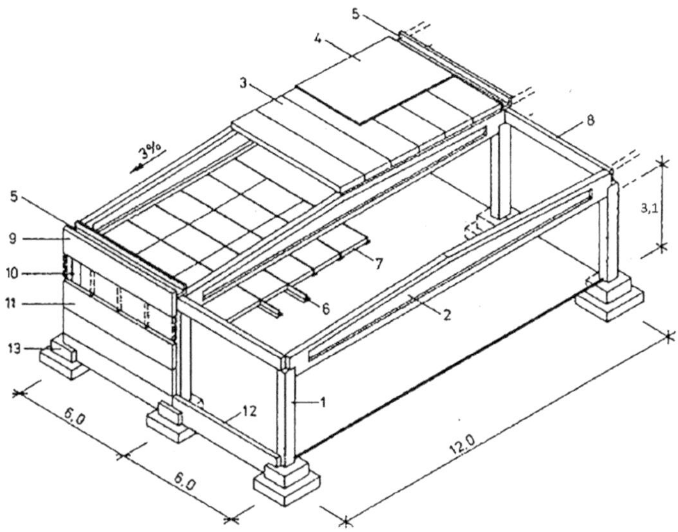

Fig. 5.2 Structural and materials characteristics of construction system C1b (Sýkora et al., 1992). Note: 1, column 0.6 m × 0.3 m; 2, prestressed girders; 3, roof panel (width 1.2 m); 4, asphalt cover on the levelling cement screed; 5, gutter element (width 0.6 m); 6, beam ceiling; 7, thermal insulating ceiling panel; 8, longitudinal stiffener 0.3 m × 0.6 m; 9, attic panel; 10, window belt; 11, wall iron–concrete panel (width 0.25 m); 12, plinth beam 0.25 m × 0.6 m; 13, prefabricated foundation foot

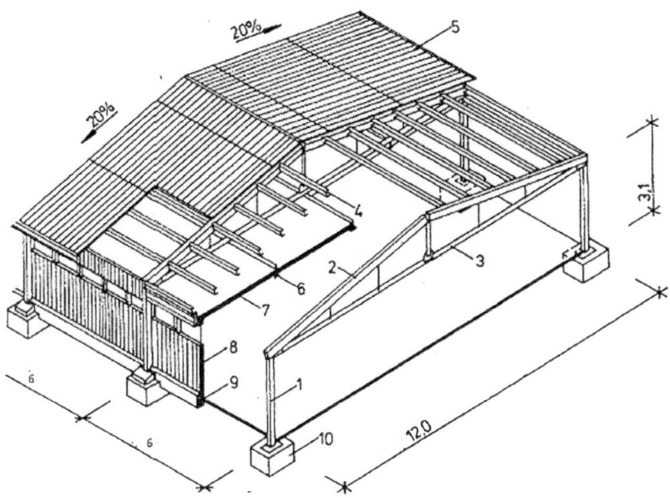

Fig. 5.3 Structural and materials characteristics of construction system S1a (Sýkora et al., 1992). Note: 1, column of rectangular cross-section welded from thin-walled profiles U 162 mm × 55 mm × 4 mm; 2, beam of rectangular cross-section welded from two thin-walled profiles U 210 mm × 50 mm × 4 mm with attachments 90 mm × 14 mm on the flanges; 3, rod 95 mm × 4 mm; 4, beam of thin-walled profile U 162 mm × 55 mm × 4 mm; 5, covering of profiled aluminium sheet KOB 1004 × 0.8; 6, beam support welded from two profiles L 80 mm × 50 mm × 6 mm; 7, thermal insulation ceiling panel made up of aluminium profiled sheets KOB 1004 and mineral wool mats of thickness 80 mm (embedded in a plastic film); 8, wall thermal insulating panel (thickness 90 mm) (panel cover with aluminium profiled sheet); 9, reinforced concrete prefabricated element; 10, prefabricated foundation foot with anchor hooks

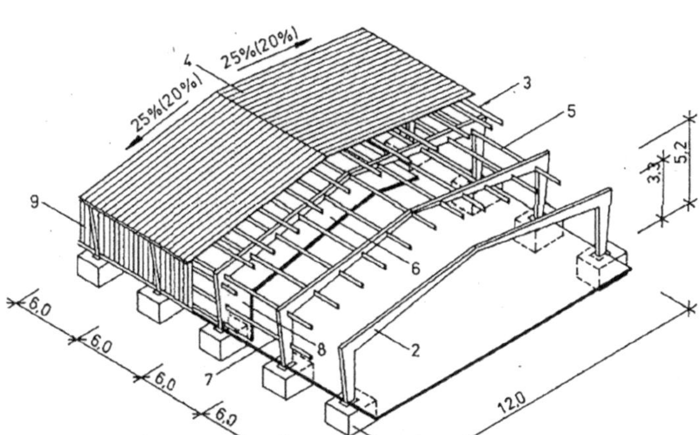

Fig. 5.4 Structural and materials characteristics of construction system S1b (Sýkora et al., 1992). Note: 1, prefabricated foundation foot with anchor holes; 2, steel frame of closed variable cross-section; 3, beams of rolled profile; 4, covering of profiled aluminium sheet; 5, ceiling beams of rolled sections; 6, ceiling panels are covered with profiled aluminium sheet with thermal insulation from mineral wool; 7, beams of rolled profiles; 8, wall panel (composition as in ceiling); 9, external cladding of profiled aluminium sheets

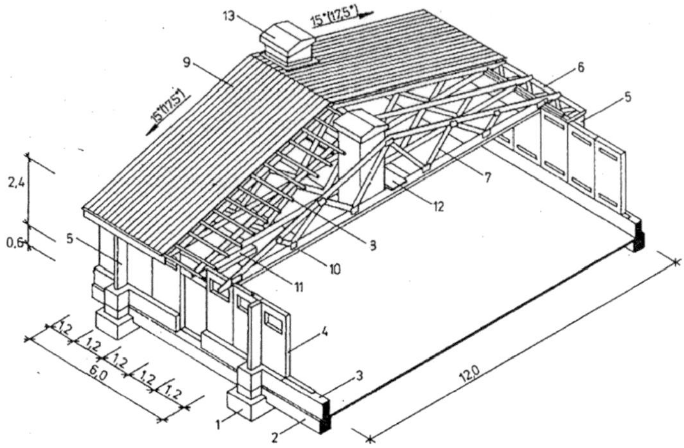

Fig. 5.5 Structural and materials characteristics of construction system W1a (Sýkora et al., 1992). Note: 1, prefabricated foundation foot; 2, prefabricated base portion; 3, masonry with thermal insulation; 4, a load-bearing peripheral panel with thermal insulation and necessary installation transitions; 5, reinforcing pillars; 6, outer beam; 7, beam girder with joint plates Gang–Nail; 8, battens; 9, covering of profiled aluminium sheet 10, ceiling thermal insulating panel

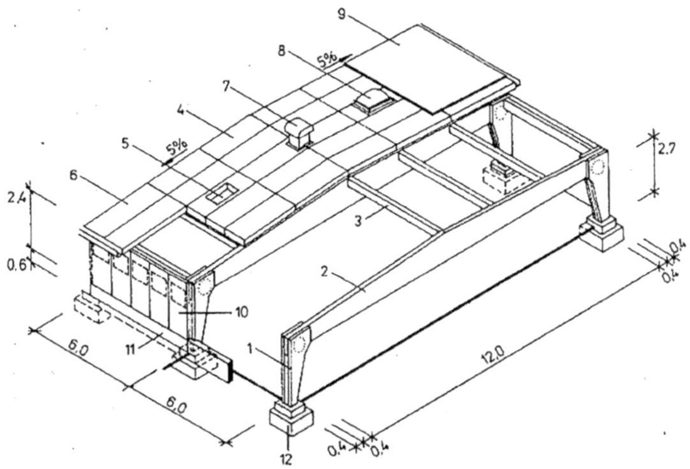

Fig. 5.6 Structural and materials characteristics of construction system W1b (Sýkora et al., 1992). Note: 1, column; 2, glued cross members; 3, glued beams 180 mm × 360 mm; 4, supporting roof thermal insulating panel with a ventilated cavity 3000 mm × 1200 mm × 270 mm; 5, roof panel with hole for skylight installation; 6, roof thermal insulating panel-gutter 4000 mm × 1200 mm × 270 mm with inlet vents to the roof; 7, double skylight; 8, ventilating attachment; 9, multi-layer asphalt roofing; 10, wall thermal insulating panel 1200 mm × 3000 mm × 100 mm; 11, masonry; 12, prefabricated foundation foot with anchor holes

was the same for all variants (iron–concrete slab, 145 mm; waterproofing, 5 mm; underlying concrete, 100 mm; compacted gravel–sand bedding, 100 mm).

The environmental characteristics of the EE, GWP, and AP of analysed structural systems or their respective materials were evaluated by the basic characteristics of the IBO database (Austrian Institute for Healthy and Ecological Building (IBO-AIBE, n.d.) Passivhaus-Bauteilkatalog). The aforementioned environmental aspects were determined according to Eqs. (5.1), (5.2), and (5.3).

$$\mathrm{EE} = \sum_{i=1}^{n} \mathrm{EE}_i \times w_i \quad [\mathrm{MJ}] \tag{5.1}$$

where

EE is embodied energy.
EE_i is coefficient of embodied energy for *i*th material [$MJ \cdot kg^{-1}$].
w_i is weight of *i*th material [kg].

$$\mathrm{GWP} = \sum_{i=1}^{n} \mathrm{ECO}_{2_i} \times w_i \quad \left[\mathrm{kgCO}_{2\mathrm{eq}}\right] \tag{5.2}$$

where

GWP is global warming potential.
ECO_{2_i} is coefficient of carbon dioxide equivalent for *i*th material [$kgCO_{2eq} \cdot kg^{-1}$].
w_i is weight of *i*th material [kg].

$$\mathrm{AP} = \sum_{i=1}^{n} \mathrm{ESO}_{2_i} \times w_i \quad \left[\mathrm{kgSO}_{2\mathrm{eq}}\right] \tag{5.3}$$

where

AP is acidification potential.
ESO_{2_i} is coefficient of sulfur dioxide equivalent for *i*th material [$kgSO_{2eq} \cdot kg^{-1}$].
w_i is weight of *i*th material [kg].

5.1.3 Specification of Research Entities

The experimental object was the construction of farm buildings designed for cattle breeding. To compare expected advantages of a modern wooden system, alternatives to the building based on the traditional building system of precast concrete and steel construction were also compared.

In the calculations of selected variants, the basic requirements placed on the construction systems were considered:

- the span of a module structure according to the applicable regulatory requirements
- assembly method of construction
- resistance to settling environment
- fire resistance according to valid legal regulations
- resistance to mechanical damage
- easy maintenance, cleaning, and disinfection

Reinforced concrete structures have considerable resistance to aggressive environments, mechanical damage, and low maintenance. Steel full-body structures have a smaller weight than concrete and more variability, but these structures place great emphasis on perfect anti-corrosion protection and the need for frequent maintenance. Modern wood-glued solid structures are resistant to aggressive environments, are quickly assembled, and provide great variability.

5.1.3.1 Reinforced Concrete Construction System C1a

Vertical support structure consists of columns attached to monolithic or prefabricated feet. The pillars are prefabricated with pockets for the storage of bundles, which are assembled from two pieces connected by a steel rod. The construction span is 12 m; the longitudinal module is 6 m. The construction height to the bottom edge of the bar is 3.15 m. Perimeter tents are made of sandwich panels (concrete–thermal insulation–concrete).

5.1.3.2 Reinforced Concrete Construction System C1b

In this single-storey construction system, the support pillars are attached to the base. Pre-filled, full-length jaws are hinged on pillars. Other structural elements include base sills, corrugated roof trusses, roof panels, and gutter beams. Construction span is 12 m with 6 m longitudinal module. The height for the housing is 3.9 m. The wall casing is made up of prefabricated sandwich panels.

5.1.3.3 Steel Construction System S1a

The construction system is in the transverse direction formed by loose columns with a hinged jaw. The construction is a three-armed, rigid drawbar with hinges. Columns and pillars are welded from thin-walled profiles. Construction span is 12 m with 6 m longitudinal module. The height of the hall is 3.6 m. The wall casing is made up of prefabricated sandwich panels with aluminium sheet covering.

5.1.3.4 Steel Construction System S1b

The supporting structure consists of a welded three-armed frame of unequal box cross-section. Frame stands are hinged on concrete feet. Construction span is 12 m with 6 m longitudinal module.

5.1.3.5 Construction System Based on Wood W1a

The supporting peripheral tents consist of thermal insulating sandwich panels, on which roof trusses are laid. The nodes are connected system Gang–Nail. The stability of the object is secured in the transverse direction by the 6 m-long snap-hooks, lined with bases and horizontal longitudinal beams. The layout of the structure is 12 m with a 6 m transverse module.

5.1.3.6 Construction System Based on Wood W1b

The main elements are full-faced glued, circumferential glued wooden pillars. The end stand is connected to the transverse (a two-armed frame is created). Intersections are embedded in construction at an axial distance of 3 m. The object's extension is 12 m with a 6 m transverse module. The height of the building is 4.8 m. The wall casing is made of thermal insulating sandwich panels.

5.2 Results and Conclusion of Analysis in the Context of Sustainability

In the context of sustainability, the underlying environmental characteristics within the defined boundaries were assessed. For better clarity, the analysis was selected according to the main technological stages (foundations, superstructure, and roofing) (Figs. 5.7, 5.8, 5.9, 5.10, 5.11, and 5.12). In addition, the individual variants were evaluated as a whole but also as part of the structure, Module 1, without a front (Figs. 5.7, 5.8, and 5.9).

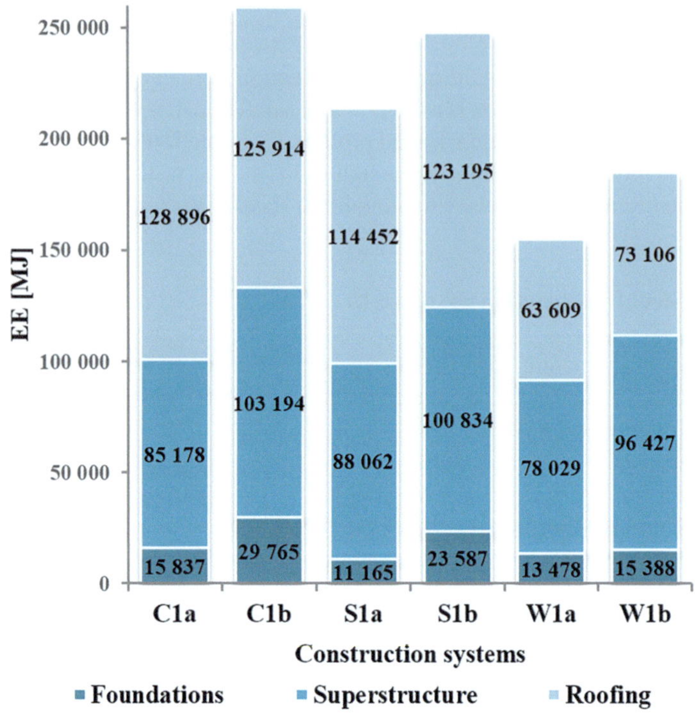

Fig. 5.7 Module 1: Analysis of EE

Analysis of selected construction systems from the perspective of EE show comparable values between prefabricated concrete structures and steel construction systems (Fig. 5.10). In the category of concrete-based construction systems, minor deviations in terms of foundation and superstructure were found for the C1b variant compared to C1a (11.4%). This difference was caused mainly by the need for more massive columns in the construction system C1b, which also involved the need for more massive foundations, which in the end resulted in higher parameters in terms of EE. Similar findings were also observed in the category of steel-based structural systems where the difference between the variants S1a and S1b was 7.6%. With these structural systems, deviations were also needed in the case of a more massive skeletal structure in the S1b system, with the need for more massive foundations. By comparing the W1a and W1b wood construction variants, a 7.5% difference was observed in favour of the W1a design system. This difference can be attributed, in particular, to the use of more massive wood elements in the W1b system with a higher weight of the upper structure and the need for a more massive foundation structure.

Comparison of constructional systems in terms of material bases in the context of embodied energy (EE) showed a minimal mean difference (4.2%) between

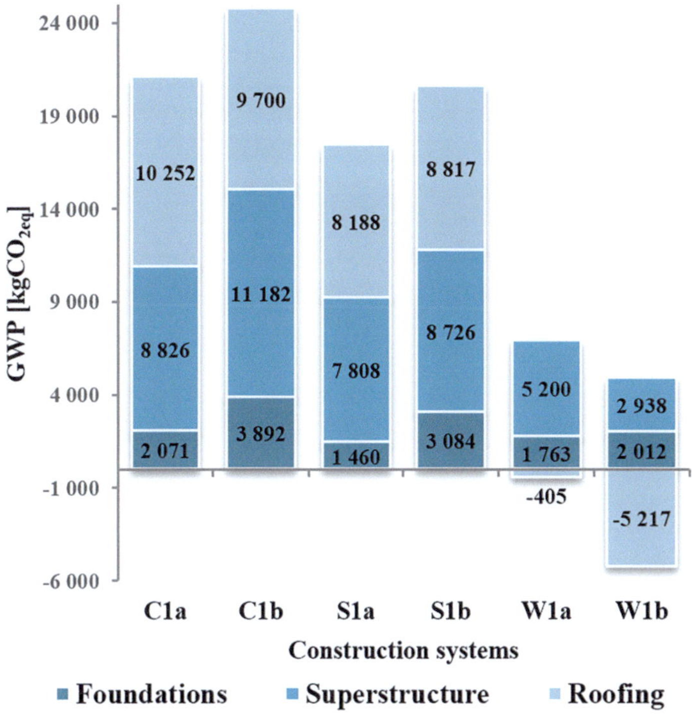

Fig. 5.8 Module 1: Analysis of GWP

concrete and steel construction systems. A significant difference was observed between steel and wood-based systems (25%) and even timber-based construction systems had a significantly more favourable balance on average of 28.6% compared to concrete-based construction systems. This finding points to the considerable benefit of the use of wooden construction systems in terms of embodied energy.

In terms of GWP, similar differences between construction variants C1a and C1b (11.7%) and between S1a and S1b (8%) were observed for EE (Fig. 5.11). A more significant difference was noted between the concrete-based construction systems and the steel-based systems (15.3%). Significantly large difference was observed for systems based on wood compared to systems based on steel (62.9–91.1%) and the difference between systems based on wood and concrete was 68.6–92.5%. This significant difference is the result that the wood has a negative GWP balance, so its use is projected in this valuation parameter. In addition, the lower weight of the wood-based construction is seamlessly reflected in the need for smaller foundation structures compared to other concrete and steel options. A significant difference was also observed between W1a and W1b timber systems (76%). This was due to the fact that the W1a system is in practice designed and raised using an aluminium-based covering to increase the undesired balance of the whole building structure

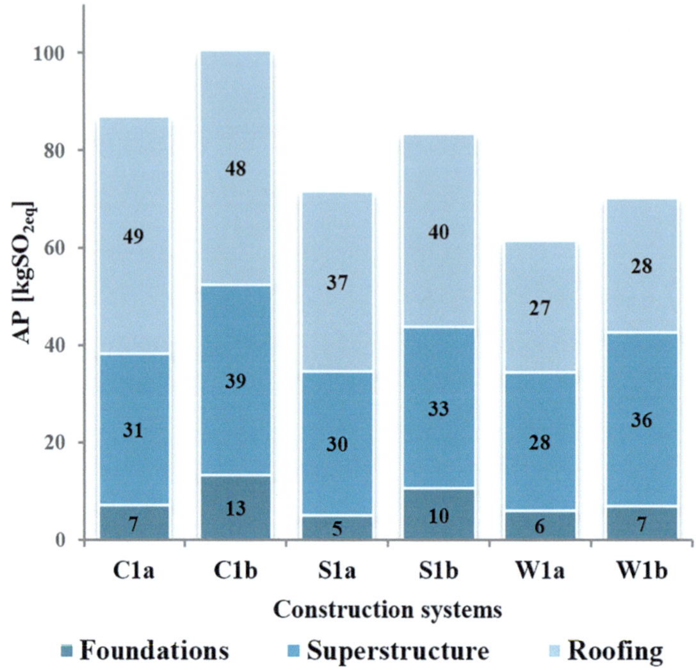

Fig. 5.9 Module 1: Analysis of AP

compared to the W1b construction where the multi-layer asphalt roofing was used. A more thoughtful and appropriate design in terms of the unsuitable balance of the materials used would achieve a more acceptable value of the resulting structure.

The assessment of construction systems in terms of AP points to 15.5–28% differences between wood-based constructions systems compared with steel and concrete-based systems (Fig. 5.12). This observed difference was not as meaningful as in the previous parameters, attributable, in particular, to the use of glued wood-based and OSB boards in wood-based system variants. Since the production of glued structural elements based on wood and OSB boards is technologically demanding and contributes significantly to emission generation, ultimately, the wood-based materials mentioned above have a better balance in terms of environmental parameters than concrete, steel, and aluminium materials. Improved wood recovery as such could be achieved by reducing and streamlining the entire production process of the production of wood-based components and in combination with materials that are effective in terms of the necessary requirements with a reasonable environmental impact.

In the category of concrete systems, a 12% difference between C1a and C1b was observed. This difference can be attributed to the more massive base structures at C1a and also to the more massive skeletal structure of the upper structure. In the category of steel-based systems, a lower difference between S1a and S1b (7.5%) was observed. Minimal differences were caused, similarly the concrete-based

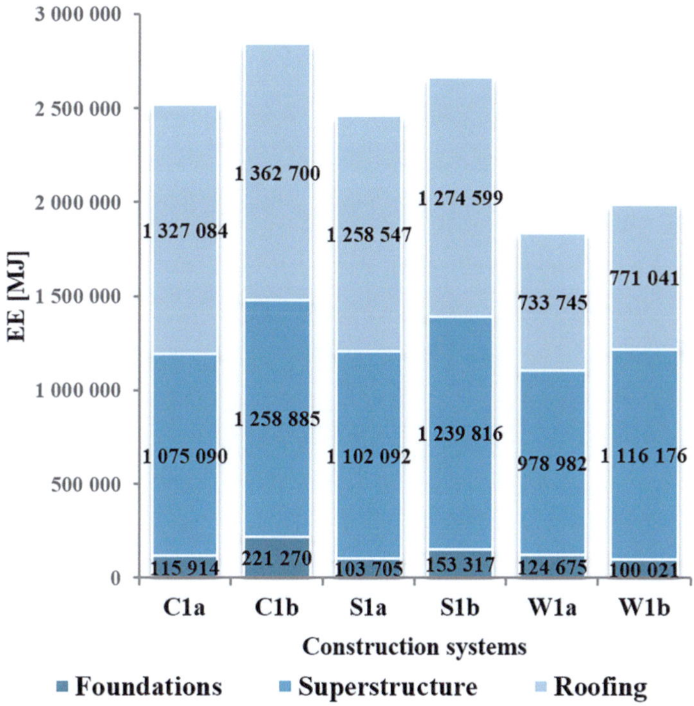

Fig. 5.10 Whole construction: Analysis of EE

systems, with more massive foundation structures and a more massive construction of the S1b skeleton compared to S1a. In terms of AP, the minimum differences in the category of wood-based systems was observed between W1a and W1b (1.8%).

The results of this analysis, which compared assembled construction systems on different material bases for primary agricultural production, point to the benefits of modern wood-based construction systems compared to traditional materials (concrete and steel). The aforementioned benefits were manifested especially in the evaluated areas of environmental parameters embodied in energy, acidification potential and, especially, global warming potential. The percentage difference of global warming potential was significantly high for steel-based systems compared to wood-based construction systems (62.9–91.1%), and the difference between the wood and concrete systems ranged from 68.6% to 92.5%. Thus, we conclude that, with a suitable combination of traditional and modern structural elements, we can achieve significantly more efficient constructions and structures with a more favourable environmental impact while maintaining the basic requirements of the bearing capacity, stability, and the resulting function of the building itself. Cleaner and more environmentally friendly technologies are of great importance for sustainability. Therefore, this analysis and its findings can contribute to a better decision-making process within a more sustainable construction sector.

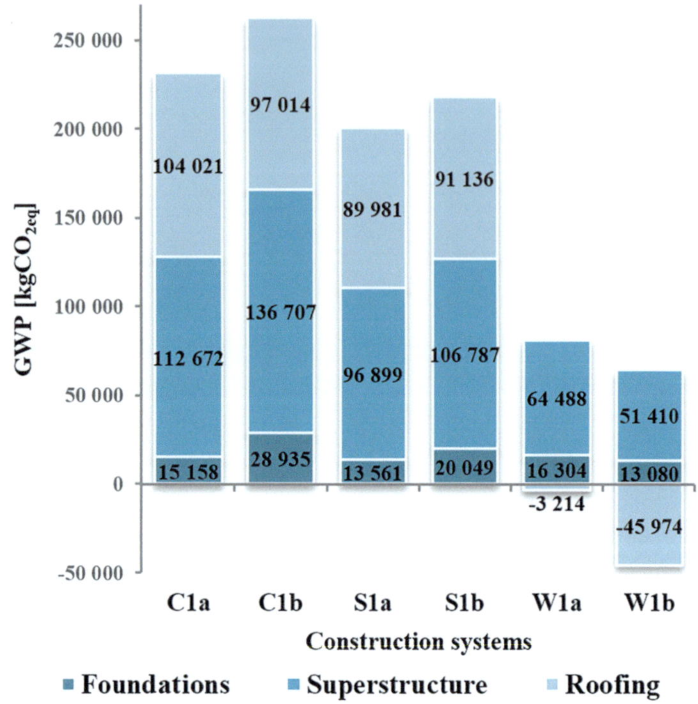

Fig. 5.11 Whole construction: Analysis of GWP

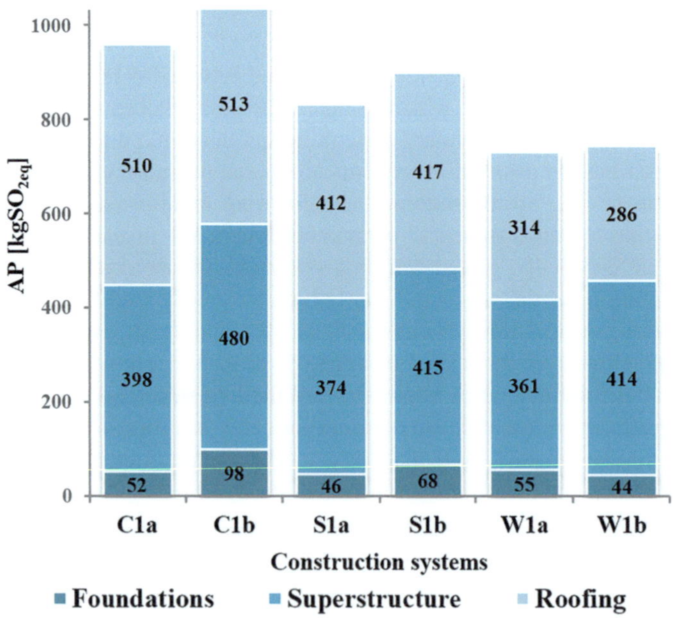

Fig. 5.12 Whole construction: Analysis of AP

Conclusion

The wood of different species of woody plants is made up of a set of plant tissues that differ in shape and function. The morphology of the tissues and their arrangement into annual rings create macroscopic features that can be observed on a cross-section with the naked eye. The macroscopic features are very important for the practical differentiation of different wood species. These properties provide essential information for the wood processor to identify the species of wood plant he is processing and to assess the physical, mechanical, and technological properties of the wood, which determine its utility value. In recent years, the Central European construction industry has been gradually returning to a more extensive use of wood. It is used not only for roof structures and the infill structures of all categories of residential buildings, but often as a load-bearing structural material or to complete the exterior and interior of various types of buildings. Modern wood constructions are based on the original wooden buildings and structures, having undergone some development to meet the current requirements and needs. In Slovakia, several types of construction systems based on wood are used, each of which has its own specific features both in terms of design and completion. This publication responds to the current trend of increasing interest in constructions based on wood and expands the knowledge in the field by introducing various contexts. The purpose of this publication is to provide an overview in the field of wood constructions, namely wood as a construction material, wood constructions in the context of sustainability, and structural systems of wood constructions.

J. Švajlenka, T. Pošiváková, *Optimizing Construction Management*,
https://doi.org/10.1007/978-3-031-84327-3

Glossary

Half-timbered construction A half-timbered construction is a construction that consists of a load-bearing wooden frame with infill. The frame consists of sill beams (horizontal beams laid on the substructure, wall, or ceiling of a lower storey), wall beams (horizontal beams where the masonry ends), vertical posts (vertical beams) mortised into sill beams and wall beams, as well as diagonal braces reinforcing the wall. The spaces created in the wall by posts and braces may be filled with wattle, but most commonly with brick or stone masonry. The surfaces of the infill were treated by plastering or rendering and most commonly whitewashed.

Log A log is a tree trunk that has had its branches and bark removed to make raw wood of tree length, not physically cut into shorter pieces. It is intended for further mechanical, chemical, etc. processing (raw material for the wood processing industry). Logs, especially from coniferous trees (pine, fir, larch, spruce), are used, for example, in the construction industry for struts, columns, piles, or for the construction of wooden log houses. The basic quantitative and qualitative properties of logs are laid down in the relevant standards.

Straining beam A straining beam is a part of wooden trusses used for their transverse reinforcement. It consists of a pair of planks with a typical cross-section of 80 × 160 mm, which (in a queen post truss) are fitted horizontally, with bolted joints, just above the wall beam on both sides of the rafter, strut, and post. In the upper part of the truss, additional straining beams may be fitted just below the centre wall beams to both paired rafters and both posts.

Acoustic properties Acoustic properties refer to the ability of a structure to prevent the propagation of sound and mechanical vibrations. Sound is a wave that propagates through space. It is the intensity of sound, stated in decibels (dB), that is important.

Agglomerated wood Agglomerated wood is a building material produced by mechanical or chemical treatment of less valuable wood (wood fibre, chips) and wood waste (sawdust, shavings). It is bonded together using waterproof adhesives

J. Švajlenka, T. Pošiváková, *Optimizing Construction Management*,
https://doi.org/10.1007/978-3-031-84327-3

to form boards, for example, fibreboard, particleboard, sawdust board. It is used to make furniture, cladding, partitions, and for other construction applications.

Airtightness Airtightness is the ability of a structure to absorb sound waves. Its value depends on the bulk density of materials and the thickness of the materials used.

Bulk density Bulk density is the mass per unit volume of a substance, including voids and pores.

Cellulose Cellulose (pulp) is a plant polysaccharide that is part of the walls of plant cells. A macromolecule of cellulose contains more than 1000 beta-D-glucopyranose building units. Unlike starch and glycogen, it features a B-glycosidic bond. The resulting chain in the natural material is a long unbranched thread. Cellulose is the main building substance of plant cells and, together with lignin, is involved in the construction of secondary cell walls. In some substances, pure cellulose (cotton) is formed, in others, for example, wood; it is accompanied by other substances: lignin, hemicellulose, and resins. Cellulose is the most abundant biopolymer on earth, with as many as 1.5×109 tonnes produced annually.

Elasticity Elasticity is the ability of a substance to resist the action of external stress: mechanical, thermal, and moisture.

Fire resistance Fire resistance is the ability of a structure to withstand the effects of fire for a specified period of time.

Hemicellulose Hemicellulose is one of the numerous heteropolymers, such as arabinoxylans, present together with cellulose in almost all the cell walls of terrestrial plants. While cellulose is crystalline, strong and resistant to hydrolysis, hemicelluloses have a random, amorphous structure with little strength.

Impact noise resistance Impact noise resistance is the resistance of a material to the spread of noise caused by walking, shaking, shocks, etc.

Anisotropy In physics, anisotropy is a phenomenon in which the physical properties of a substance change according to the direction in which these properties are measured. The opposite is isotropy.

Isotropy In physics, isotropy is a phenomenon in which the physical properties of a substance are independent of the direction in which these properties are measured. The opposite is anisotropy.

Lignin Lignin is a secondary metabolite of plants and is the second most abundant organic compound on the earth's surface after cellulose. It makes up 25% of plant biomass. The highest proportion of lignin is found in the cell walls of secondary cells that form xylem vessels and tracheids. Lignin, derived from the Latin lignum, meaning wood.

Mechanical properties Mechanical properties are defined as the relationship between the mechanical stress applied to a material and the resistance it exerts to withstand the stress.

Mechanical strength Mechanical strength is the ability of a material to resist the effects of a load. Depending on the type of load applied, we distinguish among compressive, tensile, shear, torsional, and flexural strength.

Moisture content Moisture content is the water occurring in the pores and capillaries of building materials, soils (soil moisture), or their porous media, forming a water surface, or the water in the air (air moisture).

Reaction to fire Reaction to fire (flammability) is the ability of a material to ignite, burn, or glow due to the effect of a source of combustion—it expresses how materials contribute to the intensity of a fire.

Thermal properties Thermal properties are the properties of materials that influence the process of heat transfer in them and characterize their behaviour during temperature changes.

Water vapour diffusion Water vapour diffusion is the movement of water vapour through a material and depends on the degree of moisture absorption (desorption) of the material.

References

Alchian, A. A., & Demsetz, H. (1972). Production, information costs, and economic organization. *The American Economic Review, 6*, 777–795.

Allen, E., & Iano, J. (2019). *Fundamentals of building construction: Materials and methods*. John Wiley & Sons.

Alwan, Z., Jones, P., & Holgate, P. (2017). Strategic sustainable development in the UK construction industry, through the framework for strategic sustainable development, using building information modelling. *Journal of Cleaner Production, 140*, 349–358.

Andante. (2023). *Ileh—Intelligent low energy houses*. Retrieved May 01, 2023, from http://www.ileh.eu/

Antošová, N., Šťastný, P., Petro, M., & Krištofič, Š. (2021). Application of additional insulation to ETICS on surfaces with biocorrosion. *Acta Polytechnica, 61*(5), 590–600. https://doi.org/10.14311/AP.2021.61.0590

Arcari, P. (2017). Normalised, human-centric discourses of meat and animals in climate change, sustainability and food security literature. *Agriculture and Human Values, 34*(1), 69–86. https://doi.org/10.1007/s10460-016-9697-0

ASB. (2019). *Will we give multi-story wooden buildings a chance in Slovakia as well? - Dáme šancu poschodovým drevostavbám aj na Slovensku?* Retrieved January 29, 2024, from https://www.asb.sk/architektura/rodinne-domy-architektura/drevostavby/dame-sancu-poschodovym-drevostavbam-aj-na-slovensku

Association of Wood Processors of the SR - Zväz spracovateľov dreva Slovenskej republiky. (2020). Retrieved October 28, 2020, from http://www.zsdsr.sk/

Attia, S. (2018). Modern history of sustainable architecture. In *Regenerative and positive impact architecture. Springer briefs in energy*. Springer. https://doi.org/10.1007/978-3-319-66718-8_2

Azari, R., & Abbasabadi, N. (2018). Embodied energy of buildings: A review of data, methods, challenges, and research trends. *Energy and Buildings, 168*, 225–235.

Baranová, B., Demková, L., & Arvay, J. (2021). Surface-dwelling soil macrofauna and ground beetles (Coleoptera: Carabidae) of metal post-mining spoil heaps–community composition and potential risk element bioaccumulation. *Chemistry and Ecology, 37*(6), 530–551. https://doi.org/10.1080/02757540.2021.1899162

Berglund, L. A., & Burgert, I. (2018). Bioinspired wood nanotechnology for functional materials. *Advanced Materials, 30*(19), 1704285.

Bodini, A. (2012). Building a systemic environmental monitoring and indicators for sustainability: What has the ecological network approach to offer? *Ecological Indicators, 15*(1), 140–148. https://doi.org/10.1016/j.ecolind.2011.09.032

J. Švajlenka, T. Pošiváková, *Optimizing Construction Management*, https://doi.org/10.1007/978-3-031-84327-3

Bosák, R. (2016). *With the STEKO system, you can build a house in a few days - Systémom STEKO postavíte dom za pár dní*. Retrieved February 08, 2024, from https://clanky.topreality.sk/systemom-steko-postavite-dom-za-par-dni-2142.html

Brandner, R., Flatscher, G., Ringhofer, A., Schickhofer, G., & Thiel, A. (2016). Cross laminated timber (CLT): Overview and development. *European Journal of Wood and Wood Products, 74*(3), 331–351.

Brikawood. (2020). *Building a house has never been easier! With Brikawood wooden "bricks" you can build a house like Lego, without nails and screws. - Stavba domu ešte nikdy nebola jednoduchšia! S drevenými „tehlami" Brikawood postavíte dom ako z lega, bez klincov a skrutiek*. Retrieved February 08, 2024, from https://www.kreativita.info/brikawood-ekologicky-dom/

Brundtland Report, Assembly, UN General. (1987). *Report of the World Commission on Environment and Development: Our common future* (Transmitted to the General Assembly as an annex to document A/42/427).

CEDER. (2018). *Log buildings - Zrubové stavby*. Retrieved November 10, 2022, from http://www.ceder.sk/cennik.html

Chen, Y., Okudan, G. E., & Riley, D. R. (2010). Sustainable performance criteria for construction method selection in concrete buildings. *Automation in Construction, 19*(2), 235–244.

CMD. (2014). *Center of prefabricated wooden houses - Centrum montovaných drevodomov*. Retrieved November 10, 2023, from http://www.cmd-sk.sk/hrazdena.html

Coale, A. J., & Hoover, E. M. (2015). *Population growth and economic development*. Princeton University Press.

Costanza, R., Alperovitz, G., Daly, H. E., Farley, J., Franco, C., Jackson, T., Kubiszewski, I., Schor, J., & Victor, P. (2012). *Building a sustainable and desirable economy-in-society-in-nature*. Report to the United Nations for the 2012 Rio+20 Conference. DRAFT.

Cox, N. I. (2018). *Mechanical modeling of plywood/polyurethane foam structural insulated panels in transverse bending*. Doctoral dissertation, University of Alaska Anchorage.

Cramer, W., et al. (2018). Climate change and interconnected risks to sustainable development in the Mediterranean. *Nature Climate Change, 8*(11), 972–980. https://doi.org/10.1038/s41558-018-0299-2

Čuláková, M. (2013). *Katalóg stavebných konštrukcií drevostavieb*. Retrieved September 20, 2022, from https://sites.google.com/site/hodnoteniebudov/home/katalog-stavebnych-konstrukcii-drevostavieb

Čunderlík, I. (2009). *Wood structure - Štruktúra dreva*. Publisher TU Zvolen. ISBN: 978-80-228-2061-5.

Dangelico, R. M., & Devashish, P. (2010). Mainstreaming green product innovation: Why and how companies integrate environmental sustainability. *Journal of Business Ethics, 95*, 471–486. https://doi.org/10.1007/s10551-010-0434-0

Demková, L., et al. (2023). Potentially toxic elements in soil and air along an altitudinal gradient in Tatra National Park. *Journal of Geochemical Exploration, 252*, 107268. https://doi.org/10.1016/j.gexplo.2023.107268

Do, H., & Cetin, K. S. (2018). Evaluation of the causes and impact of outliers on residential building energy use prediction using inverse modeling. *Building and Environment, 138*, 194–206. https://doi.org/10.1016/j.buildenv.2018.04.039

Drevoastavby CZ. (2013). *Construction of wooden buildings - Konstrukce dřevostaveb*. Retrieved October 27, 2023, from https://www.drevoastavby.cz/drevostavby-archiv/stavba-drevostavby/konstrukce-drevostaveb/2524-prave-hrazdene-stavby-jsou-jiz-historii

Drevodomgerlach. (2018). *Wooden building systems - Drevené stavebné systémy*. Retrieved October 28, 2023, from http://www.drevodomgerlach.sk/download/drevostavby-typy.pdf

Drevostavby-car. (2023). *Small-format modules*. Retrieved May 01, 2023, from http://www.drevostavby-car.cz/konstrukcni-systemy

Drozd, W., & Leśniak, A. (2018). Ecological wall systems as an element of sustainable development—Cost issues. *Sustainability, 10*(7), 2234. https://doi.org/10.3390/su10072234

Ďubek, M., Makýš, P., Ďubek, S., Hajduchová, S., & Petro, M. (2018). The evaluation of the content of fibers in steel fiber reinforced structures and image analysis. *Journal of Civil Engineering and Management, 24*(3), 183–192. https://doi.org/10.3846/jcem.2018.1642

Ďubek, M., Makýš, P., Ďubek, S., & Petro, M. (2018). The evaluation of the content of fibers in steel fiber reinforced structures and image analysis. *Journal of civil engineering and management, 24*(3), 183–192. https://doi.org/10.3846/jcem.2018.1642

Ďubek, M., Makýš, P., Petro, M., Ellingerová, H., & Antošová, N. (2021). The development of controlled orientation of fibres in SFRC. *Materials, 14*(16), 4432. https://doi.org/10.3390/ma14164432

Espinoza, O., Trujillo, V. R., Mallo, M. F. L., & Buehlmann, U. (2016). Cross-laminated timber: Status and research needs in Europe. *BioResources, 11*(1), 281–295.

Estav. (2023). *Structural systems of wooden buildings: Light and heavy wooden frame - Konstrukční systémy dřevostaveb: Lehký a těžký dřevěný skelet*. Retrieved November 20, 2023, from https://www.estav.cz/cz/3415.konstrukcni-systemy-drevostaveb-lehky-a-tezky-dreveny-skelet

European Environment Agency (EEA). (2015). *The European environment—state and outlook 2015*. Synthesis report. Retrieved from https://www.eea.europa.eu/soer/2015

European Union. (2011). *Regulation of the European Parliament and of the Council (EU) no. 305/2011*. Retrieved January 2, 2024, from http://eurlex.europa.eu/LexUriServ/LexUriServ.do?uri=OJ:L:2011:088:0005:0043:SK:PDF

Fang, C., Liu, H., & Li, G. (2016). International progress and evaluation on interactive coupling effects between urbanization and the eco-environment. *Journal of Geographical Sciences, 26*(8), 1081–1116.

Filho, W. L., Pociovălişteanu, D. M., Borges de Brito, P. R., & Borges de Lima, I. (2018). *Towards a sustainable bioeconomy: Principles, challenges and perspectives*. Springer. https://doi.org/10.1007/978-3-319-73028-8

Flores, E. I. S., Yanez, S. J., Guzmán, C. F., García-Macías, E., & Carlos, J. 2018. *Multi-scale structural mechanics for the modelling of cross-laminated timber buildings*. Paper presented at 6th European Conference on Computational Mechanics (ECCM 6), Glasgow, Scotland.

Giannetti, B. F., Demétrio, J. C., Agostinho, F., Almeida, C. M., & Liu, G. (2018). Towards more sustainable social housing projects: Recognizing the importance of using local resources. *Building and Environment, 127*, 187–203.

Grădinaru, C. M., et al. (2020). Sustainable development of human society in terms of natural depleting resources preservation using natural renewable raw materials in a novel ecological material production. *Sustainability, 12*(7), 2651. https://doi.org/10.3390/su12072651

Hájek, P. (2007). *Sustainable building construction and its application in Central Europe - Udržitelná výstavba budov a její uplatňování ve střední Evropě*. Retrieved April 10, 2024, from http://www.casopisstavebnictvi.cz/clanek.php?detail=465

Hájek, P., Fiala, C., & Hájek, V. (2011a). *Construction of buildings, comprehensive overview - Konstrukce pozemních staveb, komplexní přehled*. ČVUT Prague.

Hájek, P., Fiala, C., & Hájek, V. (2011b). *Construction of land structures, a comprehensive overview - Konstrukce pozemních staveb, komplexní přehled*. CVUT Prague.

Hansmann, R., Mieg, H. A., & Frischknecht, P. (2012). Principal sustainability components: Empirical analysis of synergies between the three pillars of sustainability. *International Journal of Sustainable Development & World Ecology, 19*(5), 451–459. https://doi.org/10.1080/13504509.2012.696220

Hardoy, J. E., Mitlin, D., & Satterthwaite, D. (2024). *Environmental problems in third world cities*. Taylor & Francis.

Harte, A. M. (2017). Mass timber–the emergence of a modern construction material. *Journal of Structural Integrity and Maintenance, 2*(3), 121–132.

Hawken, P., Lovins, A., & Lovins, L. H. (2010). *Natural capitalism: Creating the next industrial revolution*. Rocky Mountain Institute.

Hemström, K., Mahapatra, K., & Gustavsson, L. (2011). Perceptions, attitudes and interest of Swedish architects towards the use of wood frames in multi-storey buildings. *Resources, Conservation and Recycling, 55*(11), 1013–1021.

Hong, J., Shen, G. Q., Li, Z., Zhang, B., & Zhang, W. (2018). Barriers to promoting prefabricated construction in China: A cost-benefit analysis. *Journal of Cleaner Production, 172*, 649–660. https://doi.org/10.1016/j.jclepro.2017.10.171

IBO - AIBE (Austrian Institute for Baubiologie and Ecology. n.d.).

Invidiata, A., Lavagna, M., & Ghisia, E. (2018). Selecting design strategies using multi-criteria decision making to improve the sustainability of buildings. *Building and Environment, 139*, 58–68. https://doi.org/10.1016/j.buildenv.2018.04.041

Isbell, F., et al. (2017). Linking the influence and dependence of people on biodiversity across scales. *Nature, 546*(7656), 65–72. https://doi.org/10.1038/nature22899

Juřenčák, V. (2019). *Advantages and disadvantages of a modular house - Výhody a nevýhody modulárneho domu.* Retrieved March 04, 2024, from https://www.ekomodular.sk/blog/vyhody-a-nevyhody-modularneho-domu/

Kamali, M., & Hewage, K. (2016). Life cycle performance of modular buildings: A critical review. *Renewable and Sustainable Energy Reviews, 62*, 1171–1183. https://doi.org/10.1016/j.rser.2016.05.031

Kamali, M., & Hewage, K. (2017). Development of performance criteria for sustainability evaluation of modular versus conventional construction methods. *Journal of Cleaner Production, 142*, 3592–3606. https://doi.org/10.1016/j.jclepro.2016.10.108

Kishawy, H. A., Hegab, H., & Saad, E. (2018). Design for sustainable manufacturing: Approach, implementation, and assessment. *Sustainability, 10*, 3604. https://doi.org/10.3390/su10103604

Kleskeň, P. (2019). *Building system made of wood - STEKO, quality from Switzerland - Stavebný systém z dreva—STEKO, kvalita zo Švajčiarska.* Retrieved February 08, 2024, from https://www.stavajsnami.sk/stavebny-system-z-dreva-steko-kvalita-zo-svajciarska/

Knut, M., & NES BAU. (2020). Retrieved September 10, 2023, from http://www.nesbau.sk/

Kobl, J. (2011). *Wooden constructions: Systems of load-bearing structures, perimeter cladding - Dřevostavby: Systémy nosných konstrukcí, obvodové pláště.* Publishing Grada. ISBN: 978-80-247-4071-3.

Koc, B., & Ceylan, M. (2012). The effects of social-economic status of consumers on purchasing, behaving and attitude to food products case study of Van, Turkey. *British Food Journal, 114*(4–5), 728–742. https://doi.org/10.1108/00070701211230006

Kohout, J., Muller, P., & Tobek, A. (1996). *Carpentry, tradition from today's perspective - Tesařství, Tradice z pohledu dneška.* Grada Publishing. ISBN: 8071694134.

Kolb, J. (2008). *Wooden buildings - Dřevostavby.* Grada Publishing. ISBN: 978-80-247-2275-7.

Koželouh, B. (2004a). *Wooden structures according to Eurocode 5—Step 1. Design and construction materials - Dřevěné konstrukce podle Eurokódu 5—Step 1. Navrhování a konstrukcní materiály.*

Koželouh, B. (2004b). *Wooden structures according to Eurocode 5, Step 2. Designing details and supporting systems. - Dřevěné konstrukce podle Eurokódu 5, Step 2. Navrhování detailú a nosných systémú.* ISBN: 80–86–769-13-5.

Kristjansdottir, T. F., Houlihan-Wiberg, A., Andresen, I., Georges, L., Heeren, N., Good, C. S., & Brattebø, H. (2018). Is a net life cycle balance for energy and materials achievable for a zero emission single-family building in Norway? *Energy and Buildings, 168*, 457–469. https://doi.org/10.1016/j.enbuild.2018.02.046

Kujanová, K. (2010). *Wooden buildings, Do we know everything about them?.—Drevostavby, Vieme o nich všetko?* Retrieved January 10, 2024, from http://www.domabyvanie.eu/page/index.php?log=clanok&uid=44

Kuklík, P. (2005). *Wooden construction - Dřevěné konstrukce.* ČVUT Prague.

Kuklík, P. (2012). *The past and present of wooden buildings - Minulosť a súčasnosť drevostavieb.* Retrieved August 20, 2023, from https://www.asb.sk/architektura/rodinne-domy-architektura/drevostavby/minulost-asucasnost-drevostavieb

Kuzma, I. (2017). *Wood? A sensible decision - Drevo? Rozumné rozhodnutie*. Retrieved January 10, 2024, from http://www.fordom.sk/files/File/DREVO_ROZUMNE_ROZHODNUTIE.pdf

Langford, F. M., & Stott, A. W. (2012). Culled early or culled late: Economic decisions and risks to welfare in dairy cows. *Animal Welfare, 21*, 41–55. https://doi.org/10.7120/096272812X13345905673647

Lehocký, F. (2018). *Wooden houses, Prefab houses on a turnkey basis, "how to choose a contractor", experience directly from practice - Drevodomy, Montované domy na kľúč - „ako vybrať realizátora" - skúsenosti priamo z praxe*.

Lin, K. W., & Shih, C. M. (2018). The comparative analysis of neighborhood sustainability assessment tool. *Environment and Planning B: Urban Analytics and City Science, 45*(1), 90–105. https://doi.org/10.1177/0265813516667299

Liška, L. (2012). *History of panel wood buildings - Historie panelových dřevostaveb*. Retrieved October 05, 2023, from https://docplayer.cz/4883923-Historie-panelovych-drevostaveb.html

Lu, H. R., Hanandeh, A. E., Gilbert, B., & Bailleres, H. (2017). A comparative life cycle assessment (LCA) of alternative material for Australian building construction. *MATEC Web of Conferences, 120*, 02013. https://doi.org/10.1051/matecconf/20171200

Lu, Y., et al. (2018). Major threats of pollution and climate change to global coastal ecosystems and enhanced management for sustainability. *Environmental Pollution, 239*, 670–680. https://doi.org/10.1016/j.envpol.2018.04.016

Lukacs, I., Björnfot, A., & Tomasi, R. (2019). Strength and stiffness of cross-laminated timber (CLT) shear walls: State-of-the-art of analytical approaches. *Engineering Structures, 178*, 136–147.

Mahmoud, S., Zayed, T., & Fahmy, M. (2019). Development of sustainability assessment tool for existing buildings. *Sustainable Cities and Society, 44*, 99–119.

Malhi, Y., et al. (2020). Climate change and ecosystems: Threats, opportunities and solutions. *Philosophical Transactions of the Royal Society B, 375*(1794), 20190104. https://doi.org/10.1098/rstb.2019.0104

Marique, A.-F., & Rossi, B. (2018). Cradle-to-grave life-cycle assessment within the built environment: Comparison between the refurbishment and the complete reconstruction of an office building in Belgium. *Journal of Environmental Management, 224*, 396–405. https://doi.org/10.1016/j.jenvman.2018.02.055

Meng, Q., Chen, W., & Hao, H. (2018). Vulnerability analyses of structural insulated panels with OSB skins strengthened by basalt fiber cloth subjected to windborne debris impact. *International Journal of Structural Stability and Dynamics, 18*(06), 1850088.

Mojprojekt. (2018). *Log buildings, history of log buildings - Zrubové stavby, história zrubových stavieb*. Retrieved January 27, 2024, from https://www.mojprojekt.sk/historia-zrubovych-stavieb/

Moustafa, K. (2016). Food and sustainability challenges under climate changes. *Science and Engineering Ethics, 22*(6), 1831–1836. https://doi.org/10.1007/s11948-015-9737-y

Murillo, M., Tutikian, B. F., Ortolan, V., Oliveira, M. L., Sampaio, C. H., & Gómez, L. (2019). Fire resistance performance of concrete-PVC panels with polyvinyl chloride (PVC) stay in place (SIP) formwork. *Journal of Materials Research and Technology, 8*(5), 4094–4107.

Nemet, A., Klemeš, J. J., Duić, N., & Yan, J. (2016). Improving sustainability development in energy planning and optimisation. *Applied Energy, 184*, 1241–1245. https://doi.org/10.1016/j.apenergy.2016.10.111

Nilsen, A. (2021). *Vernacular buildings and urban social practice: Wood and people in early modern Swedish society*. Archaeopress Publishing Ltd.

Nižetić, S., et al. (2019). Smart technologies for promotion of energy efficiency, utilization of sustainable resources and waste management. *Journal of Cleaner Production, 231*, 565–591. https://doi.org/10.1016/j.jclepro.2019.04.397

National Strategy of Sustainable Development SR (NSSD SR). (2001). Retrieved from http://www.minzp.sk/files/dokumenty/strategicke-dokumenty/narodna-strategia-trvalo-udrzatelneho-rozvoja-slovenskej-republiky-cast-1.rtf

National Strategy for Sustainable Development for the Slovak Republic (NSSD SR). (2002). *Decree no. 532/2002*. Retrieved January 14, 2024, from https://www.epi.sk/zz/2002-532#cast2

Opp, C., Ziebolz, B., & Groll, M. (2018). Ecological, socio-economic and demographic analyses as prerequisites for sewage treatment problem solutions in rural areas. The case study of Dirlammen, Vogelsberg, Germany. *DIE ERDE—Journal of the Geographical Society of Berlin, 149*(2–3), 184–197. https://doi.org/10.12854/erde-2018-369

Østergaard, N., Thorsted, L., Miraglia, S., Birkved, M., Rasmussen, F. N., Birgisdóttir, H., Kalbar, P., & Georgiadis, S. (2018). Data driven quantification of the temporal scope of building LCAs. *Procedia CIRP, 69*, 224–229. https://doi.org/10.1016/j.procir.2017.11.057

Palfiová, A. (2008). *A house made of wood, a healthy and ecological alternative - Dom z dreva, zdravá a ekologická alternatíva* (Vol. 1). House and housing - Dom a bývanie. ISSN 1337-4028.

Plati, C. (2019). Sustainability factors in pavement materials, design, and preservation strategies: A literature review. *Construction and Building Materials, 211*, 539–555. https://doi.org/10.1016/j.conbuildmat.2019.03.242

Popovski, M., & Gavric, I. (2016). Performance of a 2-story CLT house subjected to lateral loads. *Journal of Structural Engineering, 142*(4), E4015006.

Quesada-Pineda, H., Smith, R., & Berger, G. (2018). *Drivers and barriers of cross-laminated timber (clt) production and commercialization: A case of study of Western Europe's clt industry* (pp. 29–38). BioProducts Business.

Ramage, M. H., Burridge, H., Busse-Wicher, M., Fereday, G., Reynolds, T., Shah, D. U., & Allwood, J. (2017). The wood from the trees: The use of timber in construction. *Renewable and Sustainable Energy Reviews, 68*, 333–359.

Ruuska, A., & Häkkinen, T. (2016). Efficiency in the delivery of multi-story timber buildings. *Energy Procedia, 96*, 190–201.

Salgot, M., & Montserrat, F. (2018). Wastewater treatment and water reuse. *Current Opinion in Environmental Science & Health, 2*, 64–74. https://doi.org/10.1016/j.coesh.2018.03.005

Sandanus, J. (2010). *Wooden supporting structure for multi-storey houses. - Drevená nosná konštrukcia pre viacpodlažné domy*. Retrieved January 29, 2024, from https://www.asb.sk/architektura/rodinne-domy-architektura/drevostavby/drevena-nosna-konstrukcia-pre-viacpodlazne-domy

Sanye-Mengual, E., Orsini, F., & Gianquinto, G. (2018). Revisiting the sustainability concept of urban food production from a stakeholders' perspective. *Sustainability, 10*(7), 2175. https://doi.org/10.3390/su10072175

Schlattmann, A., et al. (2022). Ecological sustainability assessment of water distribution for the maintenance of ecosystems, their services and biodiversity. *Environmental Management, 70*(2), 329–349. https://doi.org/10.1007/s00267-022-01662-3

Singh, R. L., & Pradeep, K. S. (2017). Global environmental problems. In *Principles and applications of environmental biotechnology for a sustainable future* (pp. 13–41). Springer. https://doi.org/10.1007/978-981-10-1866-4_2

Singh, S. K., et al. (2019a). Environmental ethics, environmental performance, and competitive advantage: Role of environmental training. *Technological Forecasting and Social Change, 146*, 203–211. https://doi.org/10.1016/j.techfore.2019.05.032

Singh, T., Page, D., & Simpson, I. (2019b). Manufactured structural timber building materials and their durability. *Construction and Building Materials, 217*, 84–92.

Smith, P., et al. (2016). Global change pressures on soils from land use and management. *Global Change Biology, 22*(3), 1008–1028. https://doi.org/10.1111/gcb.13068

Stasiak-Betlejewska, R. (2016). Innovative wooden energy efficient houses constructions. *Czasopismo Techniczne, Mechanika Zeszyt 3-M (10) 2016*, 47–52.

Staudt, F., et al. (2021). The sustainability of beach nourishments: A review of nourishment and environmental monitoring practice. *Journal of Coastal Conservation, 25*, 1–24. https://doi.org/10.1007/s11852-021-00801-y

Štefko, J. (2012). *Wooden buildings from the point of view of sustainable construction - Drevostavby z pohľadu trvalo udržateľnej výstavby*. Retrieved January 10, 2024 from http://www.google.sk/url?sa=t&=j&qFwww.prolignum.sk

Štefko, J., Balent, V., & ISOVER. (2012). *Multi-comfort buildings made of wood - Multikomfortné stavby z dreva*. Retrieved January 10, 2024, from http://www.isover.sk/repository/download/310

Štefko, J., & Reinprecht, L. (2004). *Wooden buildings, construction, protection and maintenance - Dřevěné stavby, konstrukce, ochrana a údržba*. Publishing Jaga Group. ISBN: 8088905958.

Štefko, J., Reinprecht, L., & Kuklik, P. (2009). *Wooden buildings, construction, protection and maintenance - Dřevěné stavby, konstrukce, ochrana a údržba*. Publishing Jaga. ISBN: 978-80-8076-080-9.

Štefko, J., et al. (2010). *Modern wooden buildings - Moderné drevostavby*. Publisher ANTAR. ISBN: 80-967718-9-2.

Stevens, J. R., Newton, R. W., Tlusty, M., & Little, D. C. (2018). The rise of aquaculture by-products: Increasing food production, value, and sustainability through strategic utilisation. *Marine Policy, 90*, 115–124. https://doi.org/10.1016/j.marpol.2017.12.027

Sýkora, J., Košatka, B., & Daneš, K. (1992). *Hospodářske stavby*. ARCH.

Takabayashi, H., Kado, K., & Hirasawa, G. (2019). Versatile robotic wood processing based on analysis of parts processing of Japanese traditional wooden buildings. In *Robotic fabrication in architecture, art and design 2018* (pp. 221–231). Springer.

Tejchman, J. (2014). Evaluation of strength, deformability and failure mode of composite structural insulated panels. *Materials & Design (1980–2015), 54*, 1068–1082.

Tsai, C. Y., & Chang, A. S. (2012). Framework for developing construction sustainability items: The example of highway design. *Journal of Cleaner Production, 20*, 127–136. https://doi.org/10.1016/j.jclepro.2011.08.009

Tulevech, S. M., Hage, D. J., Jorgensen, S. K., Guensler, C. L., Himmler, R., & Gheewala, S. H. (2018). Life cycle assessment: A multi-scenario case study of a low-energy industrial building in Thailand. *Energy and Buildings, 168*, 191–200. https://doi.org/10.1016/j.enbuild.2018.03.011

Varga, T. (2024). *Wood as a structural and expressive material in Slovak architecture - Drevo ako konštrukčný a výrazový materiál v slovenskej architektúre*. Retrieved January 10, 2024, from http://www.google.sk/url?sa=t&rct=j&q

Véghová, J. (2018). *How to reduce the flammability of multi-storey wooden buildings. - Ako znížiť horľavosť viacposchodových drevostavieb*. Retrieved January 29, 2024, from https://hnonline.sk/prakticke-hn/1790280-ako-znizit-horlavost-viacposchodovych-drevostavieb

Veissier, I., Butterworth, A., Bock, B., & Roe, E. (2008). European approaches to ensure good animal welfare. *Applied Animal Behaviour Science, 113*, 279–297. https://doi.org/10.1016/j.applanim.2008.01.008

Veselý, V. (2013). *Massive wooden buildings and the requirements placed on them. - Masivní dřevostavby a požadavky na ně kladené*. Retrieved January 13, 2024, from https://stavba.tzb-info.cz/vlastnostidrevostaveb/10345-masivni-drevostavby-a-pozadavky-na-ne-kladene

Wacker, J. P. (2010). Chapter 17: Use of wood in buildings and bridges. In *Wood handbook: Wood as an engineering material. Centennial edition. General technical report FPL-GTR-190* (pp. 17.1–17.13). US Dept. of Agriculture, Forest Service, Forest Products Laboratory.

Welford, R. (2016). *Corporate environmental management 1: Systems and strategies*. Routledge.

Wernerová, E., & Teichmann, M. (2017). Facility management in the operation of water supply networks. *International Multidisciplinary Scientific GeoConference: SGEM, 17*, 601–607. https://doi.org/10.5593/sgem2017/62

Wernerová, E., Endel, S., & Kutá, D. (2020). Implementation of the BIM method at the VSB-technical university of ostrava. *International Journal of Engineering Research in Africa, 47*, 133–138. https://doi.org/10.4028/www.scientific.net/JERA.47.133

Werner, F., & Richter, K. (2007). Wooden building products in comparative LCA. *The International Journal of Life Cycle Assessment, 12*(7), 470–479.
Wimmers, G. (2017). Wood: A construction material for tall buildings. *Nature Reviews Materials, 2*(12), 1–2.
Woodard, A. C., & Milner, H. R. (2016). Sustainability of timber and wood in construction. In *Sustainability of construction materials* (pp. 129–157). Woodhead Publishing.
Xu, H., et al. (2021). Ensuring effective implementation of the post-2020 global biodiversity targets. *Nature Ecology & Evolution, 5*(4), 411–418. https://doi.org/10.1038/s41559-020-01375-y
Zokaei, K., Lovins, H., Wood, A., & Hines, P. (2013). *Creating a lean and green business system: Techniques for improving profits and sustainability*. CRC press.

Index

J. Švajlenka, T. Pošiváková, *Optimizing Construction Management*, https://doi.org/10.1007/978-3-031-84327-3

MIX
Papier aus verantwortungsvollen Quellen
Paper from responsible sources
FSC® C105338

If you have any concerns about our products,
you can contact us on
ProductSafety@springernature.com

In case Publisher is established outside the EU,
the EU authorized representative is:
Springer Nature Customer Service Center GmbH
Europaplatz 3, 69115 Heidelberg, Germany

Printed by Libri Plureos GmbH
in Hamburg, Germany